EVOLUTION EXPOSED
AND
INTELLIGENT DESIGNED
EXPLAINED

BY ENGINEERING AND MATHEMATICAL PROOFS

BY

WALTER STARKEY
ENGINEERING PROFESSOR

COLUMBUS, OHIO

2011

EVOLUTION EXPOSED AND INTELLIGENT-DESIGN EXPLAINED

by

Walter L. Starkey PhD

This book is unique in that it is based exclusively on analyses and concepts of science, engineering, and mathematics, and it should be suitable and legal to be used as an auxiliary textbook in school courses on biology. This book does not include any quotation from any religious book, nor does it contain any statement that in any way could be construed as religious.

Copyright © 2011 by Walter Starkey.

Library of Congress Control Number: 2011907747
ISBN: Hardcover 978-1-4628-7225-1
Softcover 978-1-4628-7224-4
Ebook 978-1-4628-7226-8

All rights reserved. No part of this book may be reproduced or transmitted in any form or by any means, electronic or mechanical, including photocopying, recording, or by any information storage and retrieval system, without permission in writing from the copyright owner.

This book was printed in the United States of America.

To order additional copies of this book, contact:
Xlibris Corporation
1-888-795-4274
www.Xlibris.com
Orders@Xlibris.com
95962

CONTENTS

CHAPTER	PAGE
1. Introduction	1
2. My Lifelong Interest in Animals and Machines	3
3. Single-Cell Animals on Earth or Mars	6
4. Definitions Of Earth, Machines, and Animals	9
5. The Natural Forces of the Earth	12
6. The Capabilities Required to Design Machines	15
7. Who Designed the Machines and Animals?	17
8. The Intelligence of Each Designing Entity	21
9. The Reasons Behind Evolutionist's Beliefs	23
10. The DNA Molecule	29
11. DNA and Mutations	32
12. Proof #1. The Gamete Cell Must be Found	36
13. Proof #2. DNA Atoms Can Not be Rearranged by Chance	38
14. Proof #3. There are No Partially-Developed New Features	41
15. Proof #4. The Fossils do Not Support The Evolutionist's Concept of Gradual Multi-Step Evolution	42
16. The Basic Nature of Chance	52
17. Alleged Scientific Evidences for Evolution	58
18. The Family Tree of Animals	59
19. Identical Early Embryos	64
20. ARchaeopteryx, the Missing Link	67
21. The Evolution of the Horse	71
22. The Four-Winged Fruit Flies.	75
23. Structural Similarity of Animal Bones	77
24. The Miller-Urey Experiment	80
25. The Peppered Moths	82
26. The Finches of the Galapagos	86
27. Facts and Assumptions	88
28. Introduction to Intelligent Design	93
29. Intelligent Design is Philosophical Logic	101
30. Summary and Conclusions	102
Bibliography	104

CHAPTER 2. MY LIFELONG INTEREST IN ANIMALS AND MACHINES

My Interest in Animals

Starting when I was two years old, and continuing throughout my life, I have had and intense interest in, and fascination with, the animals of the earth. Also starting when I was very young, and continuing throughout my life, I have had an intense interest in machines. I can vividly remember the birth of my sister, and it is a known fact that I was two years old when my sister was born. Our family then lived in the high Sierra Nevada mountains above Placerville, California. My father was a civil engineer working on the construction of the Eldorado hydroelectric facility which used falling water from the high mountains to produce electricity for California. I remember the animals which lived in the mountains, especially the mountain lions. The animals fascinated me because I recognized the excellence of their designs, and I was captivated by their clever abilities to move about, find food, hide from enemies, defend themselves and reproduce.

My parents noted my fascination with animals, and when I was eight years old, as a Christmas present, they gave me a subscription to a magazine called *Nature*. This was a magazine about animals. Later in life I made a movie of the small animals of the Rocky Mountains, and another movie of the large animals.

When I was a junior in high-school I took a course in Zoology, and in that course I learned a lot more about animals. But the teacher of the course spent a large percentage of the course teaching us that the origins of all of the animals was by a process called evolution. He taught that every animal evolved from some previous animal, and he supported his teachings by citing alleged scientific evidences which proved that it was evolution that produced all of the animals. As a young student, being taught by a highly respected teacher, I accepted his teachings, and I became an evolutionist, but with some reservations.

After I graduated from high-school, I spent the summer working in my cousin's shop near Eau Claire, Wisconsin, where he specialized in welding and the repair of machinery, especially for the farmers who lived in the area. Cousin Earl was one of the most intelligent and highly-educated persons I have ever known. He was a walking encyclopedia. But Cousin Earl told me that the animals of the earth were designed and constructed by some superhuman being. He convincingly refuted the evidences of the evolutionists and gave me many reasons to believe that his opinions were the truth. So, by the time I was 17 years old, I had learned much about the debate between the evolutionists and those who believed that our animals were created by some superhuman intelligent designer. I then spent much of my time during the rest of my life attempting, objectively, to learn the truth about this debate.

My Interest in Machines

Let's now consider in more detail my intense interest in machinery. When I was three years old I was fascinated by the trucks and earth-moving machines that were being used to construct water reservoirs in the high mountains. I was also greatly intrigued by my father's new Chevrolet automobile. So one day when I was three years old, and when we lived in the high mountains, I climbed into my father's car and I moved the gear-shift lever, as I had seen my father do when he was driving. But moving this lever eliminated the braking effect that the gearbox was providing, and the car started rolling down the mountain with me inside of it. It gained significant speed, but it then crashed into a tree. I was thrown up and forward and my head hit one of the side-curtain fasteners which was on a windshield structure and the one-inch-long fastener entered my skull. It produced a deep hole in my head. I can clearly remember going to a medical facility where they put ice on my wound. That depression in my skull has remained there the rest of my life. It can be felt today. But this experience did not reduce my interest in machines. It was fun driving the car down the mountain, until it hit the tree.

When I was eight years old I took the wheels off of my sister's baby-buggy and put them on a 2" by 12" plank which was 12 feet long. I used it as wagon pedaling it on the side walk. But soon a policeman told be to get that thing off of the sidewalk, and never put it on the sidewalk again.

When I was nine years old my brother received, as a Christmas present, an erector set. This consisted of a wide variety of stamped metal shapes, which, together with the set's bolts, nuts, axles, wheels, gears, shafts and an electric motor, could be used to design and construct real machines. My brother, who later became a philosopher, had no interest in machines, so I adopted the erector set as my own. I spent all of my spare time for months designing and constructing machines using the parts from this erector set.

When I became 16 years old I spent much of my time chasing Model-T Ford cars on my bicycle. If I caught one of the Fords I would ask the driver if it was for sale. I finally found a 1926 two-door Model-T which I negotiated to buy for $10. It would not run. I then approached my parents in an effort to have them let me buy this car. My parents had noted my interest in machinery, and they thought that I would learn a lot by trying to get the Ford to run. So we towed the car home, and, by 6 AM the next morning I succeeded in getting the engine to run, but in the process I wore all of the skin off of the palm of my right hand, by turning the crank. I soon made a very fine automobile out of that Model-T, and I did, indeed, learn a lot about machines. During the time that I owned that Ford I removed and repaired, as needed, every part of that vehicle. In 1939 our family moved from Chicago to Louisville, Kentucky, and I drove the Model-T to the new home in one day, a distance of more than 300 miles.

After graduating from high-school, I had to select a major subject to study in college. After much discussion with my parents I decided to study mechanical engineering, because it was the profession which specialized in designing machines. So I enrolled at the University of Louisville in mechanical engineering. Several years later, after graduating from the University in mechanical engineering, I was hired by the University to be a faculty member, and I specialized in teaching the students about machine design. After teaching at the University of Louisville for four years I decided to work toward a master's degree, and possibly a doctor's degree in machine design. I was offered a scholarship by The Ohio State University, and I accepted their offer. By 1950 I had earned both the MSc degree and the PhD degree. And, incidentally, all of my grades for these degrees were A's.

I was then offered a position on the faculty of Ohio State and I became a professor at OSU, where I served for the next 30 years. I was in charge of machine design at Ohio State for all of that 30-year period.

During my years as a professor at Ohio State I continued my interest in both machines and in animals, and I continued to study the theory of evolution. I also studied what has become known in recent years as the theory of intelligent design. The remainder of this book will reveal to you what I learned about these subjects.

CHAPTER 3. SINGLE-CELL ANIMALS ON EARTH OR MARS

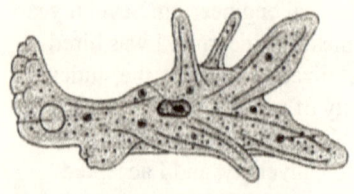

Did the Single-cell Animal Come First?

Most of this book will be devoted to the evolutionist's proposition that every species of animal evolved from some predecessor species. But this concept, of course, does not apply to the first single-cell animals, because it had no predecessor. So, to be timewise logical, we will discuss the single-cell animal before we study the other animals which, according to evolutionists, evolved one from another.

Historical Concepts of Single-cell Animal

Since evolutionists believe that all living things came into existence through the actions of natural phenomena operating strictly by chance, they, accordingly, believe that the first single-cell animal came into existence by chance. Since all animals consist largely of water, they theorize that, many years ago, some unusual event, such as a stroke of lightning, impacted upon some water-based group of chemicals, and out of this combination of circumstances a single-cell animal was born. For many years this seemed plausible, because, to evolutionists, no other theory seemed of offer greater merit, and for three other reasons:

(1) Until recently, a single-cell animal was thought to consist largely of a simple inactive substance called cytoplasm.
(2) A single-cell animal was thought to consist of very simple life-supporting features, lacking any extensive complexity.
(3) Scientists were unaware of the existence of the DNA molecule and its complex and dominant role in the life and development of animals, including the single-cell animal.

But, in recent years, Michael Behe (Behe 1) and others have discovered that cytoplasm is a myth, and that single-cell animals are extremely complex, comparable in complexity to a large array of many machines, structures and chemical facilities.

The Probability that the First DNA Molecule Arose by Chance

In later Chapters of this book we will discuss the DNA molecule which is in every cell of every animal. This DNA molecule consists of thousands of atoms of the chemicals, Carbon, Nitrogen, Oxygen, Hydrogen, and Phosphorous, known as C, N, O, H, and P. But for each animal these atoms must be arranged in a particular geometric array, and this array of atoms contains the information that directs the entire life of that cell.

It is a known fact that the DNA molecule for the E-coli bacterium, which is a single-cell animal, has 68,000,000 atoms of C, N, O, H and P in its DNA. We will assume that 68 million atoms might be typical of other single-cell animals, and we will assume that the first single-cell animal on earth needed about 68 million atoms for its DNA.

This suggests that there are several obvious questions that must be answered by the evolutionists. They include the following:

(1) From where did these atoms come?

(2) What caused them to move from their places of lodging in the environment to the DNA molecule of the first single-cell animal?

(3) What caused them to further move to their required specific positions in the array of atoms of the DNA molecule of the first single-cell animal?

The first question can be answered. Of the atoms needed, Hydrogen and Oxygen could have come from water. Carbon could have come from the Carbon-dioxide of the atmosphere. Nitrogen was in the air. And Phosphorus was in the earth. So the needed atoms were in the land, sea, and air.

Evolutionists cannot answer the second or the third question. It is inconceivable that 68 million of these atoms, by chance, would leave their locations in the land, sea and air, and travel to the location of some future single-cell animal. And then each atom had to move, by chance, to its particular location in the DNA array of atoms for the first single-cell animal. The probability that 68,000,000 atoms from the land sea, and air would, by chance, somehow move to their required locations in a DNA molecule is essentially zero.

Was my 1926 Truck Produced by Lightning Striking the Mesabi Range?

1926 Model TT Ford One-ton Truck

In my garage I have a 1926 Model TT Ford truck. It is made almost entirely of iron. In northern Minnesota, in the Mesabi Range, in the twenties, there was a lot of iron ore, iron oxide. Do you think that a few bolts of lightning hitting the Mesabi Range might have transformed some of that iron ore into a Model TT Ford truck? I think there is just as much chance of that happening as there is of lightning acting on some concentrated sea water and producing a single-cell animal many years ago.

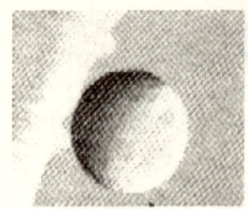
No Animal Ever Evolved by Chance on Mars

By the same reasoning we can conclude that no single-cell animal has ever evolved on Mars. The probability that complex items can be designed and constructed by chance happenings is no different on Mars than on the earth.

Unless there was some intelligent designer active on Mars, we must conclude that there are no animals on the planet Mars.

CHAPTER 4. DEFINITIONS OF EARTH, MACHINES, AND ANIMALS

The Three entities on Earth

Of the many entities that exist on the earth, the three that are of interest in this discussion are: (1) the earth itself, (2) the machines that are on the earth, and (3) the animals that are on the earth. If some visitor from another planet came to earth, these are the three things that he would at once clearly observe. In this Chapter we will define and identify what is each of these three entities, and in subsequent Chapters we will attempt to determine who or what brought each one into existence.

Definition of the Earth

The earth is defined as (1) the third planet from the sun, (2) the land surface of the world as distinguished from the oceans and the air, and (3) the dwelling place of human beings.

Definition of a Machine

A machine is defined as (1) any system, usually of rigid bodies, formed and connected to alter, transmit, and/or direct applied forces in a predetermined manner to accomplish a specific objective, such as the performance of work, (2) a machine may be a simple device, such as a lever, a pulley, or an inclined plane, that alters the magnitude or direction of an applied force, (3) a machine may be any such system or device, together with its source of energy and its auxiliary equipment, such as an automobile, and aircraft, or a jackhammer, or (4) a machine may be an intricate natural system or organism, such as the human body. All of the above are dictionary definitions of a machine.

Examples of Machines

In order to further clarify precisely what is a machine, the following list is provided which includes many of the machines that exist on the earth.

Cars and trucks	Gasoline engines	Vacuum sweepers
Trains	Steam turbines	Food mixers
Boats	Electric motors	Washers and dryers
Aircraft	Generators	Heaters and A/C
Elevators	Steam power plants	Mowers
Tractors and plows	Lathes and grinders	Bull dozers
Disks and harrows	Casting equipment	Back hoes
Seed planters	Milling machines	Front loaders

All of the above are examples of non-living machines. They all have strong and rigid parts which move. And they are activated by some source of energy, such as from the burning of gasoline, diesel-fuel, coal, or the consumption of electricity, as by electric motors. And all of these machines perform useful functions or do useful work, such as transportation, agriculture, mining, energy conversion, fabrication, earth movement, household chores, or provide for national defense

Machine Elements

Machinery can be further understood in depth by learning what are the machine elements, or the basic building blocks, of which machines are made. These include such elements as the following:

> levers, connecting rods, pistons, crankshafts, machine frames, bearings, gears, shafts, brakes, clutches, springs, hydraulic cylinders, pipes, pressure vessels, pumps, belts, pulleys, cams, roller chains, wheels, nuts, bolts, rivets, weldments, chains, cables, ropes, flywheels, fasteners, turbine blades, propellers, airfoils, lubricators, control systems, optical devices, auditory devices, heating, ventilating and air-conditioning equipment, and others.

The Figure below shows examples of several of the machine elements.

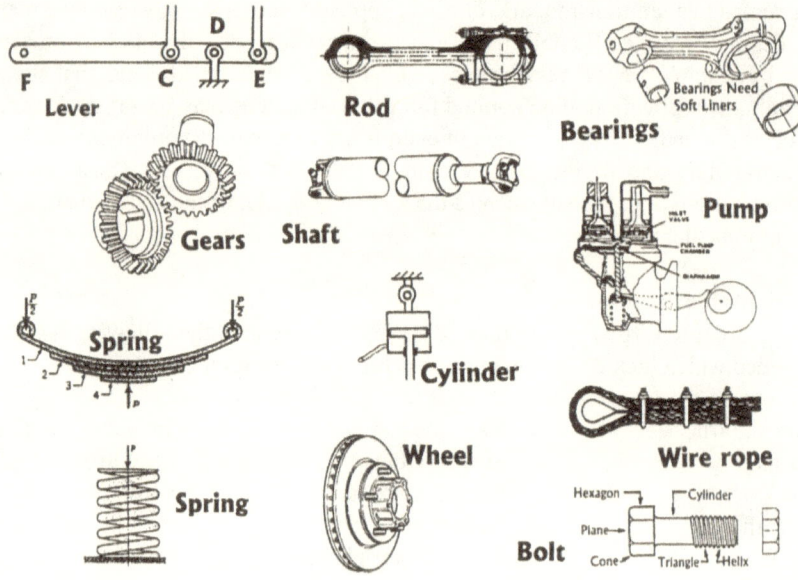

Definition of an Animal

An animal may be defined as (1) any organism which is a living individual of the biological kingdom of Animalia, (2) animals may be distinguished from plants by observing certain typical characteristics such as locomotion, fixed structure, limited growth, and nonphotosynthetic metabolism.

In order to further clarify precisely what is an animal, the following list is provided which includes many of the animals that exist on the earth.

Protozoa	Spiders	Newts
Sponges	Insects	Snakes
Hydra	Crabs	Dinosaurs
Corals	Sea worms	Turtles
Jellyfish	Sea stars	Birds
Flatworms	Sand-dollars	Rodents
Roundworms	Hagfish	Bats
Pseudocoelomates	Lamprey	Cats
Clams	Fish	Whales
Octopus	Frogs	Apes
Annelida	Toads	Men

Some of the more common specific animals that are on the earth would include raccoons, squirrels, cats, dogs, foxes, coyotes, deer, elk, moose, bears, cougars, horses, cows, birds, fish, turtles, insects, spiders, etc.

As stated before, the purpose of this Chapter is to identify and provide definitions so that the reader will understand clearly what is the earth, what are machines, and what are animals.

CHAPTER 5. THE NATURAL FORCES OF THE EARTH

The Need for Clarification

We will now attempt to clarify what hereafter will be called, "the natural forces of the earth." Evolutionists believe that all of the animals of the earth came into existence through the actions of the natural forces of the earth. So, to understand clearly what these forces are, and how they might or might not originate entities on the earth, including animals, we must study these forces in considerable detail.

The Origin of our Earth

Scientists tell us that all of the energy, and all of the impetuses behind the forces and motions of the occupants of the earth come from the sun. Scientists also tell us what must have been the origin of the sun and the earth. This subject is covered in considerable detail in *The Cambrian Explosion* (Starkey 1), but we will describe the origin of our earth very briefly here. Scientists tell us that about 13.7 billion years ago the Big Bang occurred. The Big Bang was a sudden coming into existence of an extremely high concentration of mass and energy at some location in what is now the universe.

This explosion propelled outward mass and energy. At first most of the mass was hydrogen, one proton and one electron, but later, at certain very high temperatures, some of the energy produced nuclear fusions which then brought into existence other heavier elements. Then, after several billions of years, the forces of gravity which cause all particles of matter to be attracted to each other, began to have sufficient effect to cause the accretion of mass particles toward each other at numerous randomly distributed centers of accumulation throughout the universe. This produced the galaxies, stars, and planets.

Such accumulation, due to the absorption of kinetic energy, often produced sufficiently high temperatures to cause further nuclear fusions and the creation of other heavier elements. Finally some of these accretions formed the Milky-Way galaxy, a star which is our sun, and a planet which is our earth. Fortunately, the accretions that brought about our earth contained all of the elements of the periodic table, much water and air, and rocks which contain many minerals that are useful to mankind.

The Significance of the Word, "Natural"

We now need to learn precisely what are the natural forces of the earth, which are frequently referred to by evolutionists. We should first determine the meaning of the word, "natural." The dictionary states that natural means, (1) present in, or produced by, nature, not man-made, (2) The order, disposition, and essence of all entities composing the physical universe, (3) the primitive state of existence, untouched and uninfluenced by civilization or artificiality. Artificial means, "made by man, rather than occurring in nature." So "nature" and "natural" exclude things that are made by man.

Evolutionists add to the dictionary definitions by asserting that the word, "natural" not only eliminates anything that is made by a human being but it also eliminates anything that might be made by any superhuman being. In this book we will use the term, "the natural forces of the earth" to refer to the definition that is preferred by evolutionists.

Examples of the Natural Forces of the Earth

What, then, are examples of the natural forces of the earth? These would include: wind, rain, hail, snow, lightning, freezing, thawing, volcanic eruptions, geysers, movements of the tectonic plates beneath the oceans, earthquakes, radiation from the sun, cosmic rays, bombardments by asteroids, bombardments by meteoroids, chemical reactions in water, on the earth's surface, or in the air, etc.

Evolutionists believe that one of more of these natural forces of the earth might somehow change the structure or characteristics of an animal, and might consequently produce an instance of evolution. If, as we will show later in this book, the only way that one species could evolve into another species would be by somehow adding to, or rearranging, the array of atoms in the DNA molecule of the predecessor animal, then evolutionists must believe that somehow these natural forces of the earth must be responsible for this rearrangement of atoms. We will later in this book study and determine the statistical probability that this could be accomplished.

Evolutionists believe that one or more of these natural forces of the earth might cause a significant rearrangement of the DNA atoms in an animal, which might produce an evolutionary result. They also believe that such a rearrangement might be caused by errors in the replication of DNA molecules. And, since such errors could be caused by some action of one of the natural forces of the earth, and since such error-caused rearrangements are not in any way influenced by any intelligent being, such rearrangements should be included as a possible consequence of the actions of the natural forces of the earth.

What the Natural Forces of the Earth Actually Do

We should now identify the various things that the natural forces of the earth really do produce. These forces have changed the perfectly spherical rock that was our earth immediately after it solidified from a liquid, into our current complex of mountains, valleys, and flat-lands. These forces have produced deserts, beaches, swamps, glaciers, snow-caps at the earth's poles, and fertile soil from which all of our vegetation has arisen. Although our soil and vegetation are extremely important parts of our planet, we will not include them in this book because I have no expertise in soil or vegetation. My expertise relates to machines and animals.

The Natural Forces Take Place by Chance

One of the most important features of the natural forces of the earth is the fact that the actions and the effects of these forces take place strictly by chance. The opposite of events that take place by chance would consist of events that might take place due to the influences of intelligent beings. But, with respect to the origins of our animals, evolutionists rule out any influence that might be applied by any intelligent being. Therefore, the natural forces of the earth which evolutionists rely on to explain the origins of our animals, must be clearly understood to be based entirely on actions or effects that take place by chance, and chance alone. In later Chapters of this book we will show that the statistical probability that any complex effect might take place by chance is extremely remote.

Are the Natural Forces Capable of Designing Animals?

In this Chapter we have clearly define what are "the natural forces of the earth," and we have identified many of the changes in the earth that have been brought about by these forces. We also showed that these forces act strictly by chance. Evolutionists assert that these forces have been responsible for designing and constructing our animals. To determine the truthfulness or falsity of this claim, we should establish what capabilities are required to design and construct complex entities such as machines and animals. We should then be able to judge whether or not the natural forces of the earth have these capabilities.

CHAPTER 6. THE CAPABILITIES REQUIRED TO DESIGN MACHINES

Bold Statements from a Qualified Spokesman

I am going to boldly state what are the minimum capabilities that are required to design and construct a machine. But before I make any such definitive statement, I think that the reader should learn of my qualification to make such a statement relative to machines.

First, I have studied machines all of my life. I have taken apart and physically worked on wide variety of machines all of my life. I was at one time employed in a machinery repair shop. I was at another time employed as a machinist in a factory that manufactured machines. I have a BME degree, and MSc degree, and a PhD degree, all specializing in machine design. I was an instructor and a professor in charge of teaching machine design, for four years at the University of Louisville, and for 30 years at The Ohio State University.

Simultaneously, for 30 years I managed a research organization at Ohio State which specialized in machine design. Also, for 30 years I was a consultant to industry and the government, all involving the subject of machine design. During a 40-year period, working with attorneys and judges, I served as an expert witness in more than 150 law cases which involved problems with machinery. More than 100 judges have declared me to be an expert in the subject of machine design.

In 1971 I was awarded the Machine Design Award by the national-level American Society of Mechanical Engineers. This Award, in my field, is the equivalent of a Nobel Prize. Based on my professional background, I am without question one of the most qualified persons in the world to speak on the subject of machine design.

The Capabilities Required to Design a Machine

I will now state the capabilities that are required to design and construct a machine. First, a machine cannot be designed without an intelligent designer. But this designer must have three capabilities: (1) intelligence, (2) knowledge, and (3) craftsmanship. Our early forefathers could not produce an automobile or an aircraft. They were probably as intelligent as our engineers of today, but they lacked the knowledge to produce such machines. Also, our engineers must not only have intelligence and knowledge. but also they, or their subordinates, must also have the tools and craftsmanship required to construct complex machines.

Why do we Talk About Machines?

The reader might at this juncture question why we are devoting so much space to the design of machines, when the primary focus of this book is the subject of who designed the animals. The reason that we feature machines, as well as animals, is because machines and animals have many similar characteristics, but they also are very different from each other in important ways. Actually, animals are machines, but human beings are intelligent enough to design man-made machines but human beings are not intelligent enough to design animals. This introduces the need for there to exist a superhuman being who is intelligent enough to design the animals. We talk about machines to provide a background with respect to which the reader can better understand the animals.

CHAPTER 7. WHO DESIGNED THE MACHINES AND ANIMALS?

The Three Design-capabilities and their Products

In order to determine who or what designed and constructed the animals of the earth we need to study the three entities which have been alleged to be capable of designing and constructing. These are: (1) the natural forces of the earth, (2) human beings, and (3) a superhuman being. We must also study three categories of products which might be designed by these design-capable entities. These are: (1) the earth, with today's modifications. (2) machines, and (3) animals. Our study of machines should be very helpful to our deliberations because the animals, as we will prove shortly, are really machines.

How to Recognize a Machine

Our task now is to determine who designed and constructed the machines on the earth, and we will also want to establish how this determination should be accomplished. Let's assume that we will consider some device, and we want to determine whether or not it is a machine. We will also want to find out which of our three design-capable entities was the designer of this device. To determine whether or not the devise is a machine, it is only necessary to study the device and the definitions of a machine that are recorded in Chapter 3. If the device fits the definition of a machine, then it is a machine.

Human Beings Design Machines

I will not make a statement that is based on my qualifications as an expert in the field of machine design. The statement is that all machines were designed and constructed by human beings. I have never seen a machine that was designed by the natural forces of the earth. Machines do vary greatly with respect to complexity, and the more complex a machine is, the more certain it would be that it was designed by human beings, rather than the natural forces of the earth. Human beings are the entities that smelt iron, aluminum, copper, zinc and the other metals from their ores. And it is human beings that design systems that connect rigid bodies together with bearings to make a linkage that can perform useful work.

Examples of Antique Machines

If you travel through the wilderness where the American Indians used to live you might find a bow and an arrow. This was a machine that was not made by the natural forces of the earth. Its various parts are very different from the surrounding rocks.

I have often hiked in the Rocky Mountains and have discovered old rusty machinery. After some engineering analyses, I decided that these complex devices had not just evolved out of the soil, they obviously had been designed and constructed by human beings for the purpose of mining for metal ores. Some of the machines were for transporting the ore out of the mine, and others were rock crushers. The natural forces of the earth were not intelligent enough to design and construct these machines.

The above analyses of machines might seem to consist of statements that are obvious to any fair-minded observer. If so, let's apply the same fair-mindedness to our next topic, the animals of the earth.

Animals are Machines

Let's now recall our definitions of a machine that are stated in Chapter 4. One of the dictionary definitions stated that one example of a machine was, "An intricate natural system or organism, such as the human body." If the human body is a machine, then animals, which are physically very similar to humans, must also be machines.

We all know that definitions can be short or long, depending upon the amount of detail that is included in the definition. A somewhat longer definition, which I composed, states that a machine is an assemblage of strong and rigid parts, and sometimes gaseous or liquid fluids, connected together with joints, bearings, and fasteners, in such a way that some of the parts can move with respect to the others; and, when energy is applied to the system, one or more of the parts will move and/or exert forces which perform useful functions or do useful work. Chapter 6 of *The Cambrian Explosion* (Starkey 2) devotes six pages to proving that animals are machines.

Machines are made up of combinations of the machine elements that are identified and pictured in Chapter 4 of this book. An automobile, for instance, involves many of the these elements; actually it involves most of them. Furthermore, many machines also involve, usually as sub-systems, products and processes designed by other engineers. An automobile engine involves chemical reactions, the burning of fuels; and electrical devices, such as for ignition. But, the majority of the parts of an automobile are the machine elements of the mechanical engineer, and an automobile involves rigid parts that move, and its function is a mechanical function, ie. to transport people or freight Therefore it is basically a mechanical device, and it is designed by a machine designer who is a mechanical engineer.

Are Animals Machines?

Are animals machines? Obviously they are! Animals have strong and rigid parts. The bones are fastened together by bearings and joints. The parts move with respect to each other. An animal can transport itself. It can haul freight. It is powered by energy it absorbs. It processes and burns its food. Many animals contain levers, bearings, springs, cylinders, pumps, pipes, pressure vessels, cables, lubricators, control systems, optical devices, auditory devices, and others. The bones and skeletal parts of animals are strong and rigid parts. Limbs, such as arms and legs, move with respect to a frame. Bones are often connected to each other by bearings. The bearings are cushioned by spring materials, and are lubricated by lubricating fluids. The heart is a pump, which pumps blood, a liquid. Blood flows through pipes, the blood vessels. Each lung is also a pump. It pumps air. Body ligaments are really cables. A temperature-control system is present in many animals. In humans, the body temperature stays close to 98.6 degrees Fahrenheit. Eyes are optical devices. Ears are auditory devices. The sweating phenomenon is an air-conditioning system.

Animals perform many of the major mechanical functions, which shows that they are machines. They transport themselves, and they haul freight. They haul building materials from which they build their homes. Birds build nests in trees. Beavers build houses in lakes and ponds. Bears dig holes under trees. Animals are agricultural machines. They harvest crops for their food. They engage in mining. They dig holes and move rocks. They transform energy. Their muscles transform chemical and electrical energy into mechanical work. Their hearts and lungs pump blood and air. Their blood vessels are mechanical pipes. They perform household tasks. They remove garbage and wastes, and they process food for their young. They wash themselves and their young. Many animals are prolific earth movers. Some construct tunnels; others dig burros; some construct mounds. All animals provide for self defense. Some cooperate in group warfare.

Animals are Machines
Are animals really machines? Indeed they are! In the first place, they are machines by definition. Secondly, it should be apparent from the in-depth study presented above that the characteristics of machines and the distinctive features of animals provide an excellent match. And, finally, I am an expert in the field of machinery, and I say that animals are machines.

However, in later pages of this book we will frequently use the word machine and the word animal. And, although an animal is a machine, to avoid always having to use the words man-made machine to distinguish a machine from an animal, we will simply use the word machine to identify a man-made machine and the word animal to identify an animal. The context will usually suggest what we mean by these words.

Animals are more Complex than Man-made Machines
What then is the difference between animals and man-made machines? The primary difference is that animals are almost infinitely more complex. In this section, when we use the word, "machine" it will refer to a man-made machine. Do you know of any machine that can heal itself if it is damaged? Do you know of any machine that can send electrical messages along a nerve that is made of soft animal flesh and contains no copper or aluminum wires? Do you know of any machine that has a brain that consists of a computer which is far more complex than any computer made by man? Do you know of any machine that can give birth to an offspring?

I will now make a crucially important and definitive statement. This statement will be based on my qualifications as an expert in the field of machine design. This statement asserts that there is no human being on earth who has enough intelligence, knowledge and craftsmanship to design and construct an animal. This includes all animals from a single-cell animal to a chimpanzee or an elephant. Microbiologists (Behe 1) have recently learned that a single-cell animal is as complex as a city filled with machinery and chemical plants. A single-cell animal requires about 70,000,000 atoms in its DNA molecule to manage the cell's life and its activities.

Animals were Designed by Someone having Superhuman Intelligence

The above analyses lead us to the conclusion that the animals of the earth had to have been designed and constructed by some being that is more intelligent, more knowledgeable and more capable than any human being. It would appear to me that no fair-minded person could come to any other conclusion. Because of the superior capabilities of this creator of animals, we will call this designer a super-human being.

CHAPTER 8. THE INTELLIGENCE OF EACH DESIGNING ENTITY

Intelligence Rating for Each of our Three Designing Entities

Let's now review the three entities that have been identified as possible agents that might have the capability to design and construct on earth. The table below identifies the intelligence of each of our design-capable entities.

ENTITIES POSSIBLY CAPABLE OF DESIGNING AND CONSTRUCTING

CLASS NUMBER	IDENTIFICATION OF DESIGNING ENTITY	LEVEL OF INTELLIGENCE
Class 1	The Natural Forces of the Earth	No Intelligence Whatsoever
Class 2	A Human Being	That of a Human Being
Class 3	A Superhuman Being	That of a Superhuman Being and more than that of a Human Being

A rational person should now be able to associate each of the entities found on the surface of the earth, namely, (1) the rocks of the earth, (2) our machines, and (3) our animals, with the designing entity that was really responsible for its creation. These linkages should be as follows:

> (1) The natural forces of the earth, acting solely by chance, but being governed by the laws of physics and chemistry, formed the rocks, beaches, dunes, mountains, valleys, plains, caves and other such features of our earth, but the natural forces of the earth, having no intelligence, could not design or construct a machine or an animal. But evolutionists believe that somehow these forces caused one animal species to evolve into another. Evolutionists believe that, strictly by chance, the DNA atoms of one animal became rearranged to produce another species.

> (2) Human beings designed and constructed all of the machines which are on our planet.

(3) A superhuman being designed and constructed the animals which populate our earth. No other entity has the intelligence to design our animals.

Logical Deductions

To summarize, we have now developed several unassailable facts relative to who or what designed and constructed our various entities. These entities include:

(1) The physical features of our earth.

(2) The machines which are on the surface of the earth.

(3) The animals which are on the surface of the earth.

We showed that the entity responsible for producing item (1) above, the natural forces of the earth, has no intelligence, and has no capability to design machines or animals. We showed that the entity which produces the machines was human beings, and they have considerable intelligence. We then showed that no human being is intelligent enough to design an animal, because animals are much more complex than machines. Therefore we must and did conclude that some being who has more intelligence and knowledge than that a human being must be the entity which designed and constructed the animals. We call that entity a superhuman being.

But evolutionists insist that the animals of the earth were designed and constructed by the natural forces of the earth, even though these forces have no intelligence whatsoever. I now ask the reader to be objective, fair, logical and scientific, and agree with me that the opinions and positions taken by these evolutionists are flawed, illogical, and irresponsible. Do you agree?

In the next Chapter we will explain the reasons why the evolutionists insist on retaining the beliefs that they hold on this subject, and in subsequent Chapters we will offer four analyses each of which absolutely and unequivocally proves that evolution has never occurred on this planet, and that no animal species has ever evolved to produce a different species.

CHAPTER 9. THE REASONS BEHIND EVOLUTIONIST'S BELIEFS

Three Reasons Behind Evolutionist's beliefs

Evolutionists believe that the natural forces of the earth designed and constructed the animals. The natural forces of the earth were identified in Chapter 5. These included wind, rain, freezing, volcanoes, tectonic-plate movements, meteoroids, etc. In order to evolve a new species the evolutionists must assert that these forces must have somehow moved the atoms in the DNA molecules of predecessor animals. Actually, these atoms could be moved merely by chance, such as by errors in DNA replications, as well as by the influence of some natural force. Also, some evolutionists observe that within any species there are substantial characteristic differences among the individuals of that species, and these evolutionists extrapolate these differences until they reason that a new species might arise by emphasizing these differences. All of the above phenomena will be included in what we call the natural forces of the earth. The primary characteristic of these forces is that they are not in any way under the influence of any intelligent being, and they have no intelligence of their own.

There are probably three primary reasons why evolutionists believe that these natural forces designed and constructed the animals. These reasons include the following:

> (Reason 1) Many evolutionists are atheists, and they do not believe that any superhuman being ever existed. Since human beings obviously cannot design animals, the atheistic evolutionists then have no option other than to believe in the natural forces of the earth, even though these forces have no intelligence.

> (Reason 2) Many evolutionists gloss over the fact that significant evolutionary changes in a predecessor animal can only come about by rearranging the DNA atoms in that animal's gamete cell, and they assume that evolution can be brought about by such environmental phenomena as adaptation to an environment, group isolations, etc.

> (Reason 3) Having glossed over the fact of Reason 2 above, evolutionists place great emphasis on the phenomenon of natural selection, assuming that after some vaguely defined force might have caused a change in an animal, natural selection would become active and would insure that any new feature that is beneficial to the animal would be preserved.

Reason 1: The Role of Atheism

With respect to (Reason 1) above, we need to consider the role of atheism among the adherents of the theory of evolution. The dominant role of atheism as the underlying concept upon which the theory of evolution is based is not widely understood or appreciated. As it will be shown below, most of the biologists who do research in the field of evolution are atheists. As atheists, they believe that no superhuman designer exists, or ever did exist, and therefore they are forced to believe that the animals of the world must have come into existence by the random chance actions of the natural forces of the universe.

The procedure that scientists in other fields typically use to arrive at a conclusion on a debatable topic is to consider the arguments made by both sides of the debate, and then judge which side has the most credible arguments. But on the topic of the origins of animals, the atheistic biologists cannot do this because their atheistic religion eliminates the superhuman designer as an alternative. They have no choice in the matter. Since they are atheists, they must dismiss the theory that the animals were designed by a superhuman being and embrace the theory of atheistic evolution.

Polls have proved that of the professionals in the field of biological science who specialize in studying and promoting the theory of evolution, the vast majority of them are atheists. In 1998 the elite scientific association, the National Academy of Sciences, published and distributed in public schools a book entitled, Teaching About Evolution and the Nature of Science. The primary purpose of this book was to persuade teachers to inform students that evolution is a proven fact. But in the July 23, 1998 issue of the leading science journal, *Nature*, there appeared an article which quoted a survey of all 517 biological and physical-science members of the National Academy of Sciences. Of those who responded, 72.2% admitted to being atheists, and another 20.8% were agnostics. Thus 93% of these highly respected scientists were either atheists or agnostics. In addition to the above facts, I, myself, sent a questionnaire to hundreds of biology professors, high-school teachers of biology, and curators of biological museums. I found that more than half of those who responded were atheists. I suspect that among the biology professionals who specialize in the theory of evolution close to 100 percent of them are atheists.

Why Does Biology Attract Atheists?

There are several reasons for the above statistics. First, scientists who are atheists are attracted to the field of biology and the theory of evolution because this field, more than any other, enables them to pursue both their profession, and also their religion, atheism. Secondly, departments of biology in colleges often reject any graduate student who does not believe in the theory of evolution, and many graduate students eventually become faculty. Atheists promote evolution with great zeal. Atheists are really not interested in learning about and teaching the true origins of the animals. They are interested in promoting atheism, and the theory of evolution is a convenient vehicle for achieving this goal.

However, I have known many scientists and other very intelligent and well educated people who I know are not atheists, but yet they believe that evolution is a fact. This can be explained by recognizing that scientists, in general, are held in high respect, and most people, including scientists in other fields, trust and believe the reported findings of all scientists. Hence, these trusting people accept the reported findings of evolutionists, just as they believe all scientists. They are not well informed on the subject and they do not question the statements of evolutionary biological scientists.

Ants at the South Pass on the Oregon Trail

Actually, the process of reasoning that that has been adopted by many evolutionists is similar to the reasoning of some of the ants that live on the Oregon Trail. Overlooking the 7500-foot-high South Pass on the Oregon trail near Lander, Wyoming is an anthill that has been there for thousands of years. From this overlook the ants have observed the evolution of wheeled vehicles that have passed on the road below during many years of observations.

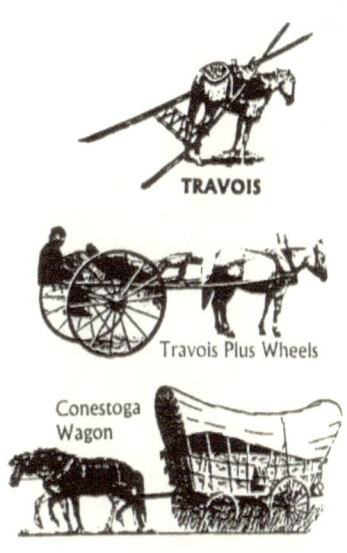

TRAVOIS

Travois Plus Wheels

Conestoga Wagon

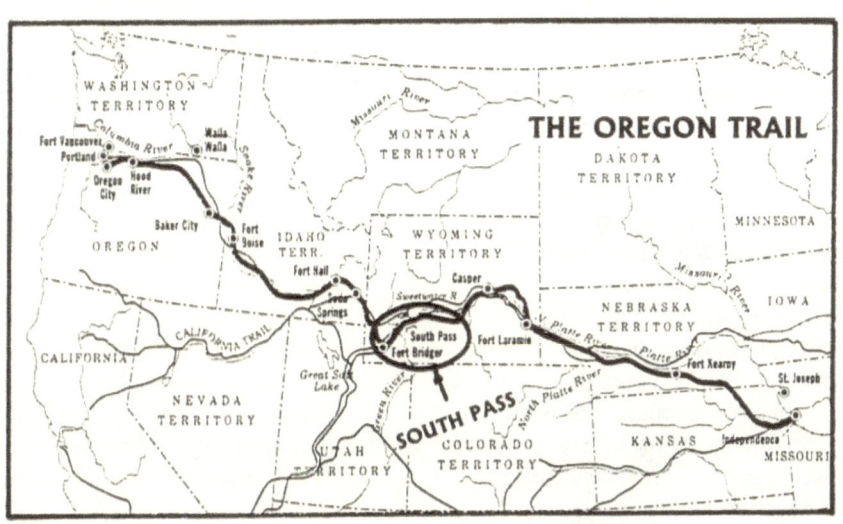

THE OREGON TRAIL

They have seen on this trail the following vehicles: A dragging Travios, a Travois with wheels, a Conestoga wagon, a 1903 Oldsmobile, a 1920 Parry, a 1931 Chevrolet, and a 1950 Ford.

The ants have frequently debated as to what was the origin of these vehicles. Some of the ants said that they had never seen nor heard of any ant that was intelligent enough to design these vehicles. Therefore they must have been designed by some Super-ant that was much more intelligent than any other ant. And they believed this even though they had never seen a Super-ant.

Other ants said that they had never seen any Super-ant being of any kind, and therefore they concluded that such a being had never existed. They believed that these vehicle must have been designed by the natural forces of the earth. These ants were called atheists, and the atheistic ants scoffed at, and looked down upon the other ants who reasoned that the vehicles must have been designed by some being who was smarter than any ant, even though they had never seen such a being.

Now we human beings know that these machines that the ants called vehicles were, in fact, designed and constructed by human beings, which were much more intelligent and capable than any ant. Today, many evolutionist's beliefs are very similar to the beliefs of one of the groups of ants. These evolutionists assert that they have never seen any Superhuman being, and therefore he has never existed.

Reason 2: Nothing Can Cause Evolution Other Than DNA Rearrangement

With respect to Reason 2 above, many evolutionists don't seem to realize that each species of animal is totally identified and determined by the array of atoms in its DNA molecule. Evolutionists speak and write endlessly about such things as morphological similarity, anatomical comparisons, lineages, descendants, family trees, identical embryos, fruit flies, horses, Archaeopteryx, similar bones, peppered moths, Galapagos finches, mutations, natural selection, Mendelian genetics, adaptation, etc. Obviously, a young student inundated with these impressive-sounding words by his respected professors would probably be overwhelmed and would likely conclude that the theory of evolution certainly must be true. But all of these words and concepts are meaningless compared to what is really the essence of the truth or fraudulence of the theory of evolution. Nothing can cause the evolution of one species into another except adding to, or rearranging, the atoms of the DNA molecule of the predecessor animal.

These environmental influences have no power whatever meaningfully to rearrange DNA atoms or to produce evolution. But evolutionists widely gloss over this fact, and they assume that some vaguely-defined influence might beneficially modify an animal, and they move quickly to the phenomenon of natural selection.

Reason 3: Misunderstanding of the Role of Natural Selection

With regard to Reason 3, natural selection, we must discuss the misplaced emphasis that evolutionists apply to this entity. Evolutionists at the highest academic level seem to misunderstand the role of natural selection as it applies to animals. Here are some quotations from college textbooks:

(1) "natural selection . . . leads to evolutionary change." (Hickman 5)

(2) The great contribution of Darwin and Wallace was that they "provided the first credible explanation for evolutionary change, the *principle of natural selection.*" (Hickman 6)

(3) "Through natural selection new species originate." (Hickman 7)

(4) "Natural selection is the guiding force of evolution." (Hickman 8)

(5) "Evolution by natural selection is now a well-documented phenomenon in nature." (Starr 3)

(6) "Natural selection is the most important microevolutionary process." (Starr 4)

(7) "Organisms . . . , what could possibly account for their diversity? . . . a key explanation is called *evolution by means of natural selection.*" (Starr 5)

(8) " . . . evolution by natural selection . . . is one of the most powerful ideas in all of science, and is the only theory that can seriously claim to unify biology." (Ridley 1)

(9) " . . . natural selection . . . is . . . responsible . . . for the whole diversification of life from a simple common ancestor . . ." (Ridley 2)

Ridley's textbook contains two Chapters on the subject of natural-selection. Chapter 5, entitled, *"The theory of natural selection"*, devotes 38 pages to this subject. High school textbooks contain similar statements

These textbook writers obviously do not understand the true role of natural selection as it applies to animals. We must here state the most significant requirements that must be met if evolution is to take place. There are two processes that must be present and active in order for evolution to take place: (1) Some mutation-causing agent must produce some new feature in the predecessor animal, and (2) natural selection must then act on the new feature and either reject it, or accept it. We have introduced the term, "new-feature-producing—agent" (NFPA) to represent the first of these two requirements, and it should be appreciated that if there is no NFPA, natural selection can have no function.

What, really, does natural-selection do? All it does is bring about extinctions. It doesn't create new features, parts, or species. Natural-selection brings about the survival of the fittest by causing the extinction of animals that are not fit. We must repeat again, every theory of evolution has to have two parts, (1) some explanation as to how new features are produced, (NFPA's), and (2) natural-selection. Natural-selection, itself, is not a New-Feature-Producing Agent and hence it, alone, cannot produce evolution. Natural-selection does, indeed, take place. But all it does is cause the extinction of animals that are not fit. Natural-selection, clearly, is not a New-Feature-Producing Agent and it cannot produce evolution. A new species can only be created by rearranging the atoms in the DNA molecule. Natural selection has no capability to rearrange atoms.

What has caused some evolutionists to seem to misunderstand this very important concept? In the first place, Darwin called his ideas the theory of evolution by natural selection. Even to this day, Darwin's proposals are referred to as Darwin's theory of natural selection. Evolutionists often tend to follow Darwin without thinking for themselves.

Paleontologist Schindewolf, on the subject of natural selection, said, "Selection by itself absolutely cannot create something new directly, but can only shape and develop what is already in existence . . . selectionism cannot explain why anything arises . . . Selection is only a negative principle, an eliminator, and as such is trivial." (Schindewolf 3)

CHAPTER 10. THE DNA MOLECULE

The DNA Molecule Serves as the Engineering Drawings for the Animal

In order fully to understand several of the proofs that will be presented in later Chapters it will be necessary for the reader to obtain a detailed understanding of the structure and the functions of DNA molecules. The DNA molecule serves a function in animals that is similar to the purposes and functions that are served by the engineering drawings for a machine. Engineering drawings describe and define a machine and they govern and manage the construction of that machine. In animals the DNA molecule defines the animal, supervises its construction and continues to control the functioning of the animal throughout its life.

What is DNA?

The term, DNA, is an abbreviation for deoxyribonucleic acid. DNA is an extremely complicated molecule, containing millions or billions of atoms. Genes are specific segments of the DNA molecule. Every cell in every animal contains a DNA molecule, and that molecule directs the structure and function of that cell. The DNA is like a book of instructions in every cell, and for the animal as a whole. The DNA molecule is different for every animal species, and it is different for every individual animal. It determines the specific structure for every part of the animal, and it manages the construction and function of every organ. It determines the personality and instincts of the animal, and it guides its growth and decline, from conception to death. What do these marvelous molecules look like?

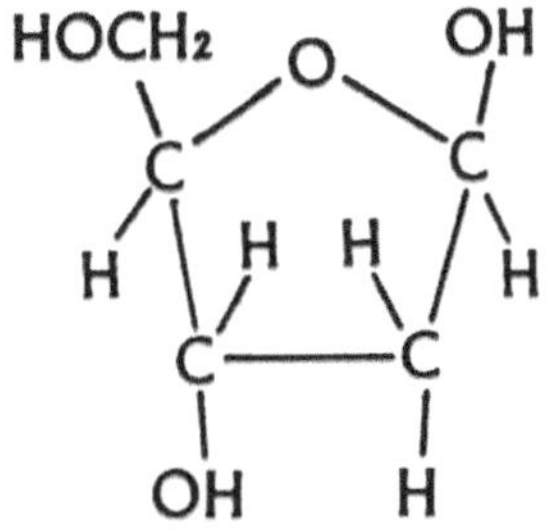

Deoxyribose Sugar (S)
Figure 10.1

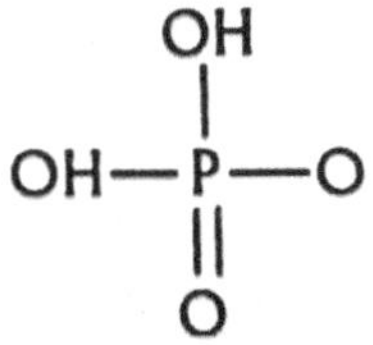

Phosphate Group (P)
Figure 10.2

The Chemistry of DNA

Every DNA molecule consists of a specific assemblage of many nucleotides. A nucleotide is an aggregation of three constituents: (1) a 5-carbon sugar group, (2) a phosphate group, and (3) a nitrogen-containing chemical base. Each of these constituents is composed of a particular group of atoms. Figure 10.1 shows the sugar group. It is called deoxyribose. We will refer to this group of hydrogen, carbon, and oxygen atoms as sugar, S. Figure 10.2 shows the phosphate group. We will call it P. It contains hydrogen, carbon, oxygen and phosphorous. There are four bases in DNA: Adenine, Guanine, Thymine, and Cytosine, which we will call, respectively, A, G, T, and C. Figures 10.3, 10.4, 10.5 and 10.6 show these nitrogen-containing bases.

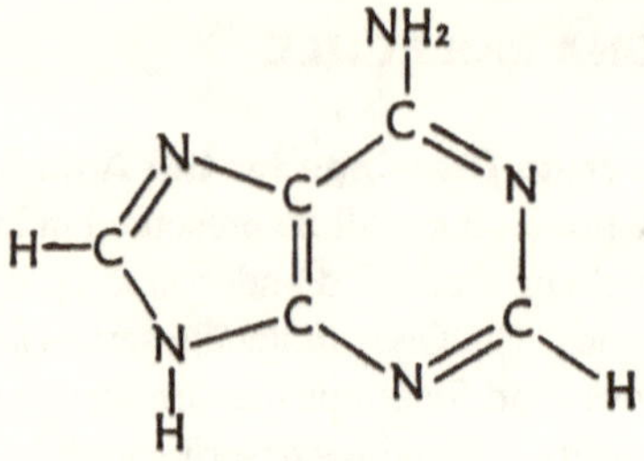

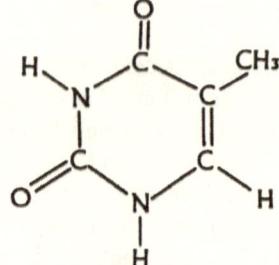

Nitrogenous Base, Adenine (A)
Figure 10.3

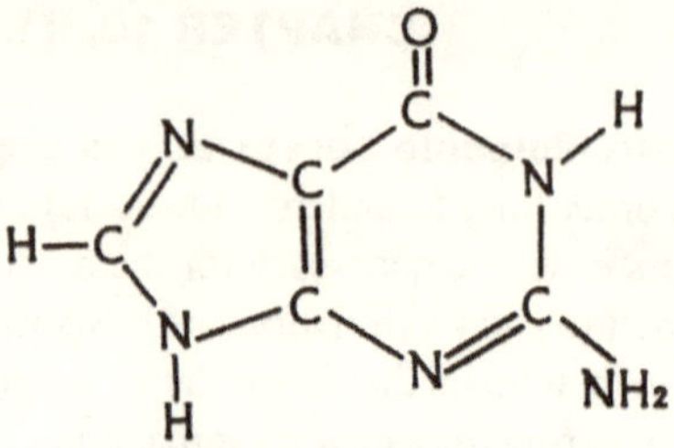

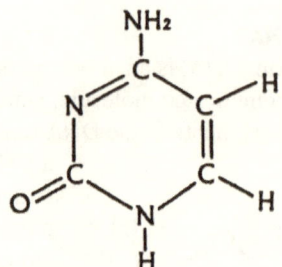

Nitrogenous Base, Guanine (G)
Figure 10.4

Nitrogenous Base,
Thymine (T)
Figure 10.5

Nitrogenous Base,
Cytosine (C)
Figure 10.6

The DNA molecule consists of a ladder-shaped assemblage of these components, arranged as shown in Figure 10.7. The ladder has two upright members and many connecting rungs. Each upright is a chain of alternating sugar groups and phosphate groups. Each sugar group is attached to one of the four bases, and the bases of the two uprights are then paired and joined together to form the rungs of the ladder. Since Adenine and Guanine are larger than Thymine and Cytosine, to ensure that all of the rungs have the same width, and hence fit well together, Adenine always pairs with Thymine, and Guanine pairs with Cytosine. Finally, the ladder structure is twisted into a double helix, as shown in Figure 10.8. The whole DNA structure may be further twisted, wound, coiled and compacted to save space.

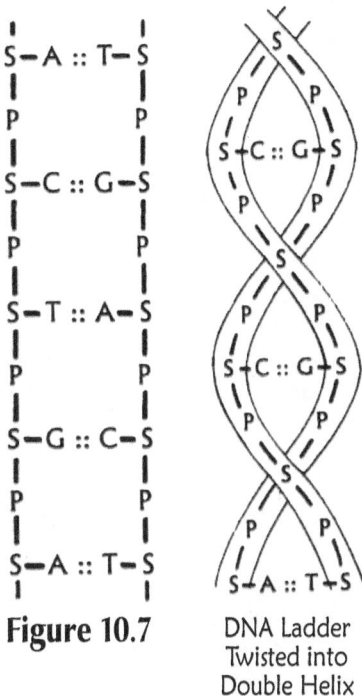

```
S—A :: T—S
|           |
P           P
|           |
S—C :: G—S
|           |
P           P
|           |
S—T :: A—S
|           |
P           P
|           |
S—G :: C—S
|           |
P           P
|           |
S—A :: T—S
|           |
```

Figure 10.7

DNA Ladder Twisted into Double Helix

Figure 10.8

One molecule of DNA for a human being contains billions of atoms, and these atoms are arranged in a pattern that is specific to each individual. The primary distinctive feature of a particular DNA molecule is the pattern of its sequence of base-pairs in the ladder. It has been estimated that the DNA molecule in a human cell, if untwisted and unwound, would be about 13 feet in length, but when twisted, coiled and packaged in a cell it may have a length of only 0.008 inches. The most significant essence of a DNA molecule is not its chemistry, but rather the fact that it is a repository of information! Its information is contained in the pattern of its almost infinitely complex arrangement of atoms, and this information is then used to direct the development and life of the specific animal which corresponds to a particular DNA molecule.

The atoms in the DNA molecule of an animal serve the same function as do the engineering drawings for the automobile. The atoms that define an animal consist of an array of atoms. But these atoms are not identical. They consist of five different elements, Hydrogen, Oxygen, Nitrogen, Carbon, and Phosphorus. These will in the future in this book be referred to, respectfully, by their elemental symbols, H, O, N, C, and P. And these atoms are placed at various specific locations on the three-dimensional, helical-shaped ladder of a DNA molecule. Then the atoms of the DNA molecule serve the function of defining the animal and controlling much of its future life.

CHAPTER 11. DNA AND MUTATIONS

A New Species Requires a New Array of DNA Atoms

Since the information which determines the species, structure, instincts and functional capabilities of each animal is contained in its DNA molecule, it should be obvious that the only way that any new species could be created from a predecessor animal would be for a new DNA molecule, having a new array of atoms, to come into existence for the new species. Even though the DNA atoms might be arranged into genes, nucleotides, promoters, enhancers, transcription factors, etc., or such items as proteins, amino acids, RNA, etc. might be involved, the basic essence of each DNA molecule is its vast geometric array of the five elemental atoms, P, H, O, N and C. A basic new array must be created for each species, and minor differences in each array will only determine the specific detailed characteristics for each individual of that species.

As we discussed in Chapter 9, in order for any evolution from one species to another to take place, both of the following two agents must be active and effective: (1) There must be some new-feature-producing agent, (NFPA), that will produce a basic change in the array of DNA atoms, and (2) natural selection must take place. Natural selection, alone, does nothing to produce a new feature. It just causes unfit animals to become extinct. Since most evolutionists believe that the natural forces of the earth are what might cause a mutation that would rearrange the DNA atoms, we will assume that the active NFPA that might produce evolution will be the natural forces of the earth.

What are Mutations?

A mutation is a sudden change in the DNA of an animal which will persist and be inherited. To understand the basic concept of how a mutation might occur, we should visualize a DNA molecule and then imagine how this molecule could be changed. Each DNA molecule of a human being contains about 3 billion pairs of nucleotides arranged in a double-helix ladder. If you count the number of atoms in a single nucleotide, based on the chemical representations of Chapter 10, you will determine that each nucleotide contains about 34 atoms. Hence, we may conclude that a single DNA molecule in a cell of a human being would contain the following number of atoms:

$$A = (3,000,000,000)(34)(2) = 204 \text{ billion atoms.}$$

The information contained in this complex of atoms constitutes the drawings and plans which determine all of the structures and functions of the cells of the individual.

Types of Mutations

Mutations that affect animals clearly do exist. There are three types of array changes in a DNA molecule which can be caused by mutations. They are the following:

(1) Atoms could be removed from the DNA of a predecessor animal.

(2) Atoms could be added to the DNA of a predecessor animal.

(3) Atoms could be rearranged within the DNA, with none removed or added.

There is no question but that mutations, due to radiation bombardments or copying errors do occur. The wings of fruit flies have been changed by exposing them to X-rays. Human beings are now believed to be subject to more than 3500 mutational disorders, but, since we have two sets of genes, these diseases rarely come to the fore.

Mutations that Cause Atoms to be Removed from DNA

If atoms were removed from the DNA of an animal, the result would be that features of the animal that were dependent upon the removed atoms would disappear, or related organs, limbs, or other body parts would be deformed or severely damaged. But this is not the desired direction of evolution. Evolution is assumed to improve body parts, add new features, and produce more complex animals. If some useful feature of an already well-developed animal were to disappear or be deformed, it would be logical that the negatively-modified animal would have less chance of survival or reproduction, and natural selection would cause that changed animal to become extinct. The removal of atoms from the DNA of an animal would not produce favorable evolutionary advancement.

Mutations that Cause Atoms to be Added to DNA

Next, let's consider mutations which might cause the adding of atoms to the DNA. The atoms that would need to be added would include carbon, hydrogen, oxygen, nitrogen, and phosphorus. Carbon is found in the air, in the form of carbon-dioxide. Nitrogen comes from the air. Hydrogen and oxygen are the components of water. And phosphorus comes from the ground. The problem is not just a matter of finding where these atoms exist on earth, but they must be removed from where they exist and be placed in the DNA molecules in just the right places. They must be so organized as to form the fantastically complex sugar groups, phosphate groups, and nitrogenous bases shown in Chapter 10, and they must be positioned in the twisted double-helix ladder in just the right positions to operate as part of the DNA molecule. Furthermore, to have any lasting effect on offspring, these mutation-causing agents would have to seek out and find the particular gamete cell that is going to produce the next progeny. Do you really believe that this transportation and positioning of atoms could occur by chance?

If Evolution Occurred There were Atom-adding Mutations

On the subject of mutations that might cause the adding of atoms to a DNA molecule, another concept is of significance. There are two ways that the animals of the earth could have come into existence: (1) every animal could have evolved from some less complex animal, starting with the first single-cell animal, and ending with the human being, or (2) every animal could have been designed and constructed from scratch, independent of all of the other animals. If the first of these two ways represents the truth, which is the opinion of most evolutionists, then it must be concluded that there had to have been mutations of the atom-adding type. And, if this were true, then what was the source of these atoms? Evolutionists must accept that they did really fly in from the environment of their own accord, by chance, and beat the almost infinite odds described above? But evolutionists cannot explain how these atoms could have been added.

A single-cell animal has about 68,000,000 atoms in its DNA molecule, whereas a human being has about 204,000,000,000 atoms in its DNA. If, as evolutionists believe, there has been a continuous process of evolution throughout the past half-billion years, involving ever more complex animals, then it is obvious that millions of mutations must have occurred of the atom-adding type. If millions of atoms were added, where did these atoms come from, what caused them to fly in from their sources to the DNA molecules, and how did they know where in these molecules they needed to be located?

On the other hand, if the second of the two ways defined above were the truth, there would be no problem, because the atoms to be added are available from scratch, and would be added by an intelligent designer.

Mutations that Cause Atoms to be Rearranged Within a DNA Molecule

Mutations of the atom-removal type discussed above do not support the concept of the evolution of animals of ever-increasing complexity. And it is very difficult for evolutionists to explain where the new atoms might come from, or how they could be transported to their proper locations, to produce the atom-added type of mutations. Therefore, in recent years, many evolutionists, by default, have decided to support the concept that evolution probably took place by the rearranging of atoms within a DNA molecule. Let's consider this in more detail.

Mutations due to the rearrangement of the DNA atoms, like the others mentioned above, could be caused by one or more of the natural forces of the earth. What, specifically, would happen if a mutation-causing agent acted upon a DNA molecule and rearranged some of the atoms of the DNA? If the agent were from sun-radiation, an X-ray, a cosmic-ray, a virus, or a chemical, the impact from the agent presumably would, by the forces of physics or chemistry, knock some of the atoms out of their normal locations, and move them to new locations.

If the mutation-causing agent consisted of errors in copying during DNA replication, the result would be the same, but instead of atoms being knocked from one location to another, they would simply be placed in different locations because of errors in copying. As an animal develops from its initial egg, new cells are constantly brought into existence, and adults generate new cells for replacements, and in both cases the DNA for each cell must be the same as the basic DNA which defines that particular animal. This is accomplished by a copying mechanism, and, if the copying were not done correctly, errors in copying may result. When the DNA of one cell is replicated, it is not only copied, but enzymes actually perform the function of proofreading the replication. Therefore, copying errors are very rare. It has been estimated by researchers that there will be only one replication error in the copying of 100,000,000 nucleotides, which would contain more than 3 billion atoms. (Starr 1). But copying errors do occur, and some evolutionists believe that these errors might cause evolutionary progress. The question that we must now consider is whether or not these copying errors, or any other ingredient of the natural forces of the earth, could produce a new species?

The Contents of This Book

The title of this book is, *Evolution Exposed and Intelligent-Design Explained*. We haven't said much yet in this book about intelligent design. The reason for this is that we are of the opinion that, in order to fully appreciate the concept of intelligent design, it is helpful if the reader is cogently informed of the fact that the alternative theory, evolution, really has no basis in fact. It is a theory based on assumptions, not facts. After one realizes that the theory of evolution is not supported by scientific evidences, and that unassailable scientific and mathematical proofs will establish that evolution never could have occurred, the reader might then prepared to embrace the theory of intelligent design.

Accordingly, most of this book will be devoted to providing a thorough education on the theory of evolution, and at the end of the book we will discuss intelligent design. The next four Chapters provide four proofs that evolution never did occur. Much of the rest of the book will refute the nine alleged scientific evidences which evolutionists typically quote to support the theory of evolution. In the final Chapter of the book we will specialize on the subject of intelligent design.

CHAPTER 12. PROOF #1. THE GAMETE CELL MUST BE FOUND

Several Unassailable Proofs that Evolution Never Occurred

In each of the next four Chapters, including this Chapter, an indisputable proof will be offered which will prove that evolution, which is alleged to cause one species to evolve into another, has never occurred on this earth. In this Chapter, it will be shown that for evolution to occur, a mutation-causing agent must find and act upon a particular cell in the body of the predecessor animal, and it will be shown that the statistical probability that this cell could be found is essentially zero.

Finding the Gamete Cell

It must be appreciated that in order for any mutation-causing agent to bring about a DNA change that will be inherited by the offspring, the DNA change can not occur in just any cell of the body of the predecessor animal. It must occur in the DNA molecule of the fertilized egg cell that would be scheduled to produce the next offspring. Such a cell will be called a gamete cell. Obviously for a mutation-causing agent to act on any other cell of the body would have no evolutionary effect because cells other than the gamete cell have nothing to do with offspring. So, the first statistical hurdle that must be overcome for evolution to proceed is for the mutation-causing agent to find the gamete cell.

In the body of a human being who weighs about 150 pounds there are about 3 trillion cells. For evolution to take place in a human being, or in another similar-sized animal, the mutation-causing agent must find the gamete cell, by chance, out of 3 trillion cells. Hence, since all aspects of evolution take place strictly by chance, there would be only one chance in 3 trillion that the agent would be successful in finding the fertilized gamete cell.

In these analyses we will find it convenient to deal with what we will call our typical animal. This should be an animal between the size of a single-cell animal and the size of a human being. We will arbitrarily define this animal to be 1/10th of the weight of a human being. If a human being has in its body 3 trillion cells, then our typical animal would have in its body, 1/10th of this, which is 300 billion cells. Hence, in our typical animal, the chance that a mutation-causing agent might find the gamete cell would be one chance in 300 billion.

In addition to the numbers problem, the mutation-causing agent would also be confronted with a time problem. The agent would have to act on the gamete cell, not at just any time, but it must act within the limited time period during which the female is in heat and mating takes place. The precise additional constraint and its effect on probability related to this time problem would be hard to estimate. But it would certainly add substantially to the already Herculean difficulty of finding the gamete cell and producing any positive evolutionary progress.

For other animals of lesser or greater weight, the number of cells in their bodies would be roughly proportional to their weights. For example, a 50-pound animal might have one trillion cells.

In conclusion, it would probably be safe to assert that, taking into consideration both the problem of finding the gamete cell among the billions of cells in a body, and the problem of finding the gamete cell shortly after it has been fertilized, the statistical probability that a mutation-causing agent would successfully find the gamete when an offspring is being originated, would be one chance in several billions or trillions. In other words, the statistical probability that the gamete cell could be found is essentially zero. Proof #1 asserts that because the gamete cell cannot be found, evolution could not even get started.

Actually, we could terminate this book at this point. The gamete cell must be found in order to produce evolution, and, based on statistics, it cannot be found. But, to be complete in our analyses, let's consider what other statistical hurdles would have to be surmounted to produce any evolution from one species to another.

CHAPTER 13. PROOF #2. DNA ATOMS CAN NOT BE REARRANGED BY CHANCE

Scientists agree that the information which determines the species of every animal is provided by the particular array of atoms which are contained in the DNA molecule for that species, and for any animal to evolve into a new species, this must be accomplished by changing the array of atoms in the DNA molecule of the predecessor animal.

Evolutionists assert that such a rearrangement of DNA atoms can be accomplished by the actions of the natural forces of the earth. They speculate that this might be accomplished by the effects of one or more of the natural forces mentioned in Chapter 4, or by errors in replication. Atoms must be added to, or rearranged in, the DNA molecule of the predecessor animal.

Now let's determine the statistical probability that such atomic movements could take place by chance.

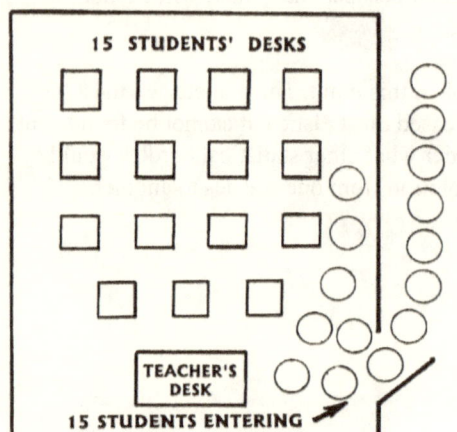

Figure 13.1 Will the 15 Students, by chance, select the seats assigned by the Teacher?

The Seating of Students

Consider first a simple analogy which illustrates the required statistical analyses. Please refer to Figure 13.1. Assume that a teacher is going to have 15 children in her class for the next school year. Before school starts she decides to assign a particular seat to each child. Now assume that she elects to have the children all come into the room at the same time and she will let the children, themselves, select their own seats. What is the statistical probability that each child will, by chance, select the seat that the teacher intended to assign to that child?

When the first student selects his seat there will be one chance in 15 that he will take the seat that the teacher had selected for him, because there would then be 15 seats from which he could choose. If the first student, by chance, did select the right seat, then there would be one chance in 14 that the next student would select the right seat, because he had 14 seats from which to choose. For the third student there would be one chance in 13 that he would select the teacher's choice for him. This procedure would then continue, involving for the remaining students one chance in 12, one chance in 11, etc. After all of the students had selected seats it should be apparent that the statistical probability that all of them would select their teacher-assigned seats would be the following:

One chance in (15)(14)(13)(12)(11)(10)(9)(8)(7)(6)(5)(4)(3)(2)(1), which equals one chance in 1.3 trillion.

Do you fully understand what this means? This means that there would only be one chance in 1.3 trillion that the required array of student seating would be accomplished by letting the students select their seats without the teacher's supervision.

Since one chance in 1.3 trillion is essentially zero probability, we can conclude that the probability of getting the students into their assigned seats by chance, just for 15 students, is essentially zero. It would not happen.

Now let me ask you this question. Suppose that the teacher, instead of letting each student select a seat by chance, she personally would usher each child to the proper seat. What do you think would be the statistical probability that she could accomplish the seating objective. I think it would be 100%, one chance in one. Why? Because instead of relying on chance, she relied on the application of information and knowledge. The teacher acted as an intelligent designer. Things that statistically cannot be accomplished by chance, can easily be accomplished by the application of intelligent design.

Finally, if you had visited the teacher's classroom just after the seating had been completed, and you saw that each student was in the right seat, which method of seating would you conclude had taken place? If you were of the opinion that the teacher didn't exist, then you would be forced to assume that the seating had been accomplished by chance. But if you saw that the seating had been accomplished, then you must conclude that the teacher did exist, and that she applied intelligent design.

The Evolution of Animals

The evolution of animals is identical in essence to the seating of the students, except that, for the case of animals, DNA atoms correspond to the students, and the correct placement of these atoms in the DNA molecule corresponds to the correct placement of the students in their seats in the room.

In the cases of the evolution of animals the number of atoms that need to be properly seated in the new array is not 15, but rather they are a substantial percentage of the numbers of atoms in the DNA molecules. These numbers, for a few animals, include, approximately, the following:

> For a single-cell animal, 68 million atoms.
> For our typical animal, 20 billion atoms.
> For a human being, 200 billion atoms.

If the statistical probability that 15 atoms could not be properly moved by chance, what do you think would be the statistical probability that millions or billions of atoms could be correctly moved by chance, in a DNA molecule? The answer is that the probability is zero! And this means that evolution, based on chance, could not ever have taken place on earth.

But maybe, in the case of the animals, some designer who had superhuman mental and physical capabilities. If this superhuman being could place the atoms where they needed to be, new animals could be created. Wouldn't it take less faith to believe in a superhuman intelligent designer than to believe that statistically impossible evolution powered only be the natural forces of the earth could design and construct our animals?

The Application of Spins

We are all familiar with politicians who apply "spins" to facts and events. If some fact or event is politically damaging to a politician he might give a spin to it and claim that it is really supportive of his positions. If you think that some partisan evolutionist might somehow spin the conclusions of this Chapter, think again. The statistical evidence that proves that evolution never did occur is so overwhelming and of such a huge magnitude that no one could possibly give it a spin or refutation that would be meaningful. We have shown that the statistical probability that just 15 DNA atoms could be successfully moved to their required locations by the chance actions of natural forces would be one chance in 1.3 trillion. But the number of atoms that would typically need to be moved for one species to evolve into another would be millions. The statistical probability that just *one* million atoms could be moved would be one chance in the following number:

$$(1,000,000)(1,000,000-1)(1,000,000-2)(1,000,000-3)(1,000,000-4)$$

etc. until the number is $(1,000,000-999,999)$.

This number would be close to infinity. So, the probability of the successful evolution of one species into another would be one chance in infinity, which is zero. We must conclude, absolutely, that evolution by chance could not, and never did, occur. These numbers, and the conclusion to which they lead, cannot be refuted or diminished by spin.

Conclusion

The proof #2 of this Chapter may be the most important concept of this book, and it will be referred to frequently in the later Chapters. This conclusion is so important that it will again be repeated here, for emphasis. If there is only one chance in 1.3 trillion that 15 atoms could be relocated, by chance, in their proper places in the DNA molecule of a successor animal, then the chance that, for a typical animal, 20 billion atoms could be properly seated by chance is zero. This means that evolution could not, and did not, ever occur.

Again, this book could end at this point, because, in each of Chapters 12 and 13, we have presented absolute proofs, mathematically, that evolution could not, and never did occur.

CHAPTER 14. PROOF #3. THERE ARE NO PARTIALLY-DEVELOPED NEW FEATURES

Would Evolution Suddenly Stop working?

If evolution were really responsible for the origins of the animals of the earth, including humans, then evolution certainly would not all of a sudden stop working. It would be taking place in the past, in the present, and in the future. Therefore, on your body you should be able to detect and see several new features that are evolving. Some would be 25% completed, some would be 50% completed, and some would be 75% completed. Surely, for all new features that are more than 50% completed you should be able to determine what the new features are going to be. Maybe an eye would develop on the end of your finger. This would be very useful. Maybe your digestive system is evolving to allow you to eat and digest grass. This would eliminate starvation for human beings. Maybe your hands and feet are developing claws. This would help you escape a wolf if you could climb a tree. Maybe you are developing the ability to spin a web, like a spider. A strong web would help you capture prey.

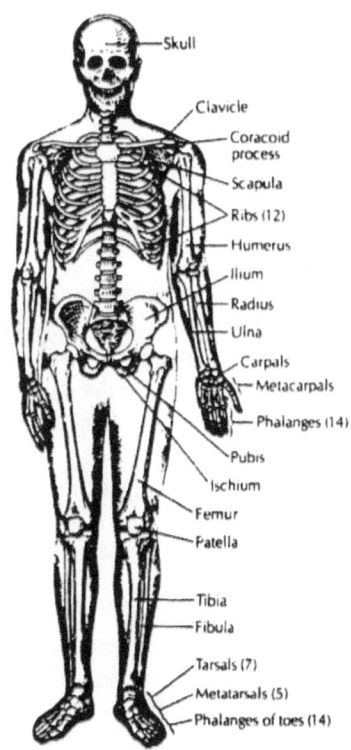

How Many New Features do You See?

Now, take off all of your clothes and look at your body and count how many developing new features you can see. I think I know how many you will see. It will be zero. None. Not one. Now perform the same study for the animals of the earth. There are about 50,000,000 species of animals on the earth. If evolution were developing new features today, wouldn't it be logical that some new features could be seen on some of the animal? Would evolution suddenly stop for all 50,000,000 of these species. But all of these animals have no partly-finished new features. Why? The answer is because evolution is not working today, and if that be true it would be logical to conclude that it has never been working.

The fact that you do not see any evidence that evolution is working now, is Proof #3 that evolution has never taken place on the earth.

CHAPTER 15. PROOF #4. THE FOSSILS DO NOT SUPPORT THE EVOLUTIONIST'S CONCEPT OF GRADUAL MULTI-STEP EVOLUTION

We Need Facts in Addition to Theories

For the purpose of orientation, let's recall again that an observer of the earth would see on its surface three entities:

(1) the rocks, hills, mountains, valleys, oceans, lakes, streams, etc. that were created by the natural forces of the earth, which had no relationship with any living being,

(2) thousands of machines that were designed and constructed by human beings, and

(3) the animals of the earth.

Evolutionists believe that the animals of the earth were designed and constructed by the chance actions of the natural forces of the earth, even though these forces have no intelligence. Evolutionists believe that the process of evolving a new species was composed of many small steps each of which consisted of rearranging the atoms in the DNA molecule of a next-offspring gamete cell of some predecessor animal.

But other scientists believe that, since no human being is intelligent enough to create an animal, and since animals are far more complex than machines, the animals must have been designed and constructed by some superhuman being who had much greater intelligence and knowledge than any human being.

To determine which of the above theories represents the truth, it would be helpful if we had some facts which are pertinent to the subject, and upon which we could rely. Fortunately, such facts do exist. They are the fossils of animals that have lived on the earth during the past 540,000,000 years.

What Are Fossils?

Animal fossils are the remains of ancient animals preserved in the earth's crust. Some examples of fossils are shown in Figure 15.1. In most cases, a fossil results if an animal is

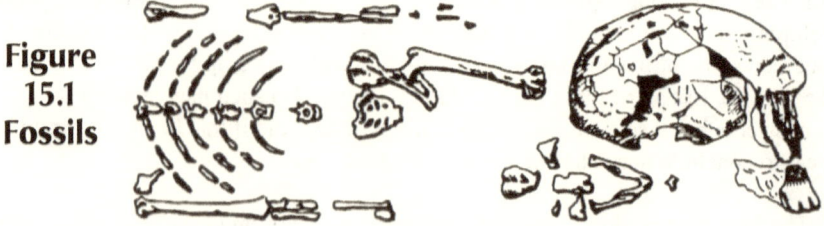

Figure 15.1 Fossils

suddenly buried, either on land or at the bottom of the sea, and then, with the passage of time, the soft parts of the animal decay and are lost, but the hard parts, such as bones, teeth, and shells, persist because dissolved minerals of the earth replace the cells of these parts, and their shapes are retained to form rock-like replicas. In some cases soft parts of buried animals can form a casting mold which eventually may leave an imprint on a rock. These fossils are found by digging in the earth, and, now, after more than a century of extensive research and recovery, "There are a hundred million fossils . . . in museums around the world." (Kier 1) "About 250,000 fossil species have been identified." (Godfrey 2)

For our study of the fossils of animals we will divide this study into two sections: (1) a study of the fossils of the animals of the Cambrian Explosion, and (2) a study of the fossils of the animals that lived after the Cambrian Explosion, during the past 540,000,000 years.

The Fossils of the Cambrian Explosion

Figure 15.2   Shale in Mountains

Many millions of years ago, at a location which is now in western Canada, at the edge of a sea, there was a high and steep earthen bank that overlooked the sea. About 540,000,000 years ago, during the Cambrian period, many tons of this bank suddenly slid into the sea. It no doubt was a huge mud slide. It happened that at the bottom of the sea there were hundreds of marine animals, including creatures of a wide variety of types and sizes. These animals were trapped by the mud, and, due to its great weight, and its mineral composition, these animals, over the years, were turned into fossils. As the years went by, the mud turned into shale, the area uplifted, and, in 1910, at Burgess Pass in the Rocky Mountains, near Field, British Columbia, these fossil animals were discovered by human beings. Figure 15.2 shows shale such as at Burgess. The mud and silt in which the animals had been buried was so fine-grained that the animals were replicated in meticulous detail. These are now the several thousand fossils of the dark gray Burgess shale, which are securely locked in the cabinets on the 2nd floor of the Smithsonian Institution in Washington, DC.

More recently, these same types of animals have been found in fossil beds in Greenland, China, Siberia and Namibia. All of the animals are of the same date, about 540 mya.

The Cambrian Explosion Produced a Unique Group of Animal Fossils.
This group of animal fossils is unique, in all the history of the earth, for several reasons. First, they are very diverse in body plan and size. Figure 15.3 shows several of these animals. They were mostly marine invertebrates, but the November, 1995, issue of *Nature* reported the finding, in China, of a 2-inch-long primitive chordate, which, by its age, would be a member of this group. These animals included bristle-worms, lamp-

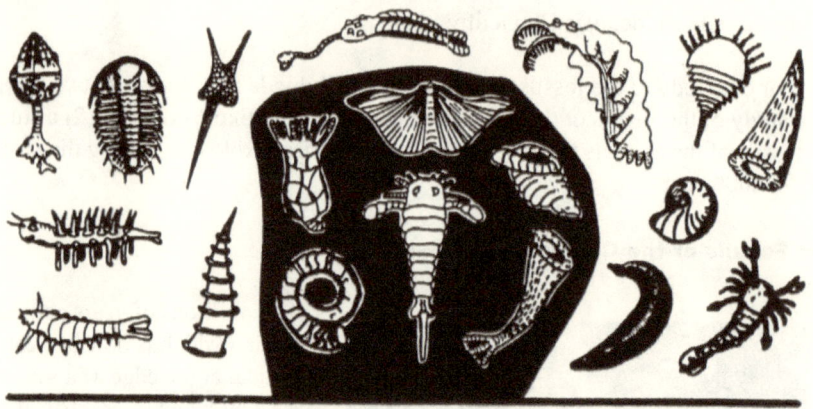

Figure 15.3 Animals From the Cambrian Explosion 543 mya.

shells, jellyfish, arthropods, ancient relatives of lobsters, crabs and spiders, trail-making and burrowing worms, starfish, etc. In fact the diversity of these animals was so great that they have been described as representing the whole animal kingdom. Scientist Conway Morris stated that in this group of animals, "nature invented the animal body plans that define . . . phyla." (Morris 1) Biologist, Cecie Starr, stated that "nearly all major animal phyla evolved during Cambrian times . . ." (Starr 8) Researcher, Michael Levine, has found that "Many new species have appeared since then. But nature drafted few, if any, new body plans after the Cambrian." (Levine 1) Biologist, Rudolf Raff, said "we've had these same old body plans for half a billion years." (Raff 1)

Secondly, these animals were advanced in design and in features. One animal, Opabinia, had five large advanced eyes, and a proboscis that resembled a fire hose. A trilobite had some 20,000 eyes, a system apparently more advanced than that of any arthropod of today. (Luria 1) Another animal was a three-foot-long shrimp-like predator that grasped its prey in circular jaws, which, mechanically, acted like a camera shutter. Another looked like a legless lobster with an elephant's proboscis that ended in a pair of claws. These animals had bodies, limbs, and internal organs of great sophistication. Stephen Gould called them "Weird Wonders". Just look at the animals of Figure 15.3. Do they look like animals that have just evolved a step or two from a protozoan?

Thirdly, and by far the most sensational attribute of these Cambrian animals, is the fact that they all appeared suddenly, almost instantaneously, and without any ancestors. The sudden and instantaneous appearance of this large group of advanced and fully-developed animals has been likened by scientists to an explosion. Hence, the sudden arrival of these animals is widely referred to as the "Cambrian Explosion."

Time Magazine Features the Cambrian Explosion

Figure 15.4

The feature article in the *Time* magazine issue of December 4, 1995, was an account of the Cambrian Explosion. They called it "Evolution's Big Bang." The information contained on the cover page of this issue is shown in Figure 15.4. In this article, author, Madeleine Nash, wrote, "Then, 543 million years ago, in the early Cambrian . . . creatures with teeth and tentacles and claws and jaws materialized with the suddenness of apparitions. In a burst of creativity like nothing before nor since, nature appears to have sketched out the blueprints for virtually the whole of the animal kingdom." (Nash 1) When did this Explosion occur? Based on new studies in the 1990's, Nash wrote, "Virtually everyone agrees that the Cambrian started almost exactly 543 million years ago." (Nash 2)

The Explosion was Really Instantaneous

We have used the word, "instantaneous" to describe the suddenness with which these animals appeared. How long did it take for this creative activity to be completed? Science editor Mark Hartwig reported that "discoveries in 1992 and 1993 have shrunk the explosion's estimated duration . . . to about 5 million years." (Hartwig 1). The earth is 4.6 billion years old. Five million years is 0.0001 (one one-hundredth of one percent) of the age of the earth. In geological time, this is, indeed, instantaneous.

The Cambrian Animal had no Predecessors

When evolutionists find fossils of dozens of new, fully-formed, advanced and complex animals, they immediately begin to search for the ancestors of the new animals. They assume that they must have evolved from some predecessors through many millions of years of evolution. But in the case of these Cambrian-Explosion animals, their ancestors cannot be found.

Mark Hartwig, stated that "in an instant of geological time, almost every animal phylum seemingly popped into existence from nowhere." (Hartwig 2)

Schindewolf wrote, "In the Cambrian rocks, we encounter for the first time . . . an abundance of well-preserved and clearly interpretable fossils." (Schindewolf 7)

Oxford evolutionary zoologist, Richard Dawkins, referring to the Cambrian-Explosion animals, stated, "It is as though they were just planted there, without any evolutionary history."

Paleontologist, Romer, wrote "Below this Cambrian period, there are vast thicknesses of sediments in which the progenitors of the Cambrian forms would be expected. But we do not find them . . ." (Romer 1)

Luria, Gould and Singer wrote, "Geologists have discovered many unaltered Pre-Cambrian sediments, and they contain no fossils of complex organisms." (Luria 2)

Zoologist Harold Coffin wrote, "If progressive evolution from simple to complex is correct, the ancestors of these full-blown living creatures in the Cambrian should be found; but they have not been found . . ." (Coffin 1)

Fossils of Animals Which Lived After the Cambrian Explosion

In the previous section of this Chapter we have been discussing the animals of the Cambrian Explosion. In this section we will study the fossils of animals which lived during the past 540,000,000 years, after the Cambrian Explosion. Evolutionists cannot explain the unique phenomenon of the sudden appearance of complex animals which appeared suddenly and instantaneously without any evidence of predecessor animals. Their theories are incapable of explaining this Explosion. But the evolutionists insist that their theories are capable of explaining the appearances of animals during the past 540,000,000 years. They assume that during this period the process of evolution must have occurred very gradually, and that each new feature of an animal must have come into existence through a series of small steps. They envision an initial small change in an animal which would add to the animal's ability to survive in its environment, and which would also enable that animal to produce offspring which would inherit the same small change. They then envision a second small change which would further enhance the capabilities to survive and produce offspring. After many such small changes, the evolutionists argue, a useful new feature might emerge, and, if enough such changes were brought about, a new species might emerge.

Besides the Fossil Record Other Facts Negate Evolution

All body units of any complexity are composed of many individual parts. They are multi-part systems. And they will not function unless all the parts are present. Such systems include arms, legs, fingers, wings, tails, fins, eyes, ears, sex organs, brains, digestive systems, cardiovascular systems, pulmonary systems, nervous systems, systems for defense, food-gathering systems, etc. If any major part of one of these systems or sub-systems is missing the system is useless.

If you believe that all animals and all body parts came into existence by a gradual process of multiple-step evolution, then describe for us the male sex organs of a mammal when they were 50% finished. Or describe the female sex organs of a mammal when they were 50% completed. Also, explain how the male sex organs, which consist of dozens of highly specialized parts, evolved on one individual, and how the completely different highly specialized female sex organs evolved on another individual of the same species. How did these sex organs function when they were 25%, 50%, or 75% completed. It is a biological fact that without the presence of all of the individual parts of the male and the female reproductive systems, the sex organs for the pair would not function at all.

Could you describe the multi-step process by which the odorous spraying system of a skunk evolved? Similar questions could be asked concerning the eyes, ears, and all of the other complex systems which are possessed by animals.

All of the parts of the body of an animal must be completed simultaneously in order for the parts to function. And this is directly contrary to the evolutionist's theory of gradual evolution by multiple steps.

Actually, the presence of a useless partially-finished new part on an animal would likely result in that animal becoming extinct.

Animals Procreate Unchanged Offspring for Millions of Years

The fossil record indicates that the general pattern of the existence of animal species is that, after their sudden appearance, they may persist without change, often for millions of years, and then become extinct. David Raup, of Chicago's Field Museum, stated, "Species appear in the sequence very suddenly, show little or no change during their existence in the record, then abruptly go out of the record." (Raup 1)

Schindewolf states that the "horseshoe crab . . . are found today in the coastal regions of . . . North America . . . This crab first appears in the Bunter (Lower Triassic) . . . This species is identical . . . to . . . recent representatives . . . The genus has a life span of . . . two hundred million years." (Schindewolf 5)

Based on fossil research performed at the British Museum of Natural History, Craig reveals that species of the following animals have lived on earth, unchanged, for the years indicated: turtles, 275 million years (my); crocodiles, 195 my; silverfish, 395 my; cockroaches, 345 my. (Craig 2)

Biologist, Professor Ridley, states, "In the fossil record, stasis—evolutionary lineages that persist for long periods without change—is common." (Ridley 3)

Swedish paleontologist Jarvik has stated that, "The main vertebrate stem groups became anatomically specialized some 400 to 500 million years ago and have changed relatively little since then." (Jarvik 1)

Also earth scientist, Steven Stanley stated, "The record now reveals that species typically survive for a hundred-thousand generations, or even a million or more, without evolving very much . . . After their origins, most species undergo little evolution before becoming extinct," (Stanley 1)

Obviously these facts concerning the longevity of species is directly in contrast with the evolutionist's theory of evolution.

Examples of Fossils that are Missing

If the contention by evolutionists were true, that evolution took place gradually and involved many small steps, then the fossils should support this contention. If all animals came into existence by the process of having mutations cause small changes in predecessor animals, then some of the new parts being evolved would involve bones, or other hard materials, that would form fossils. There should then be thousands of fossils of partly finished parts. Some would be 25% completed, some 50% completed, and some 75% completed. But, in spite of the fact that there are about 100,000,000 fossils in the various museums and laboratories of the world, the fossil record provides no evidence of the existence of any partially completed feature. On the other hand, the fossils reveal that all animal species appeared suddenly, fully developed, having no partly-completed parts.

If invertebrates evolved to become fish with a backbone, where are the fossils showing 25-percent-finished backbones? If fish evolved to become amphibians with pelvis bones and legs, where are the fossils which show that fins changed into partially-developed legs?

All fish and amphibians lay their eggs in water, and the eggs are fertilized after being laid. But reptiles lay shelled eggs on land. Fossils of eggs have been found. Reptile eggs had to be fertilized in the body of the female before the shell was formed. This required an entirely new set of sex glands and instincts. Such changes could only come about by adding thousands of new atoms to the animal's DNA. How could that happen, just by chance? Where are the fossils of the half-formed eggs?

Evolutionists believe that reptiles changed into birds. They believe that reptilian scales turned into feathers. Where are fossils of partially-formed feathers? A bird can only fly if its bones are hollow, thin-walled tubes, reinforced against elastic instability by internal braces. Are there any fossils of half-finished bird bones?

Evolutionists believe that reptiles also evolved into mammals. Mammals have three bones in their ears; reptiles have only one. Are there any fossils of intermediate animals with 1.5, 2, or 2.5 bones in their ears? Reptiles have four bones in the lower jaw; mammals have only one. Are there any intermediate fossils with 3.5, 3, 2.5, or 2 jaw bones? Are there any bones in your body which are partially-developed portions of some new feature? Are you evolving? All of these fossils of partially-developed new features are missing! These half-finished new parts are not just in short supply, they are absolutely and totally missing!

The Opinions of Experts on the Role of Fossils

If the theory of evolution were in fact the correct explanation for the source of new species, it would be expected that the fossil record would show a long series of minor changes which would gradually transform one species into another. But, the fossil record does not support this pattern of events. The best proof of this is contained in the statements of recognized scientists, which are quoted below. All of the scientists quoted below are evolutionists. But they are honest enough to state the facts even though these facts contradict their theory of evolution.

In 1950, the esteemed German paleontologist, Otto Schindewolf, wrote, "Contrary to the classic theory of evolutionary descent, the . . . designs are not smoothly connected by a long chain of transitional forms . . . but they appear in contrast with one another, set apart by large discontinuities . . . by means of sudden, discontinuous direct refashioning." (Schindewolf 2).

In 1953, paleontologist George Simpson stated, "most new species . . . appear in the record suddenly and are not led up to by known, gradual, completely continuous transitional sequences." (Simpson 1)

In 1975, geologist Arthur Boucot said, "the inability of the fossil record to produce the 'missing links' has been taken as solid evidence for disbelieving the theory." (Boucot 1)

In 1977, Paleontologist Stephen Gould said, "All paleontologists know that the fossil record contains precious little in the way of intermediate forms, transitions between major groups are characteristically abrupt." (Gould 2)

In 1979, Paleontologist David Raup, of Chicago's Field Museum, stated, "Species appear in the sequence very suddenly, show little or no change during their existence in the record, then abruptly go out of the record." (Raup 1)

Paleontologist David Raup also said, "Darwin was embarrassed by the fossil record . . . we are now about 120 years after Darwin and the knowledge of the fossil record has been greatly expanded. We now have a quarter of a million fossil species, but the situation hasn't changed much . . . we have even fewer examples of evolutionary transition than we had in Darwin's time." (Raup 2)

In 1980, in his Inaugural Lecture at the University of Queensland, geologist J. B. Waterhouse said that after 150 post-Darwin years, "the gaps have not been plugged."

In 1981, Zoologist David Kitts said, "... paleontology ... had presented ... difficulties ... the most notorious of which is the presence of 'gaps' in the fossil record. Evolution requires intermediate forms ... paleontology does not provide them." (Kitts 1)

In 1982, earth scientist, Steven Stanley stated, "After their origins, most species undergo little evolution before becoming extinct," (Stanley 1)

In 1984, Dr. Colin Patterson, Senior Paleontologist at the British Museum of Natural History, wrote to Luther Sunderland, "You say that I should at least show a photo of the fossil from which each type of organism was derived. I will lay it on the line—there is not one such fossil for which one could make a watertight argument." (Patterson 1)

In 1986, the Hickman Authors wrote, "Most major groups of animals appear abruptly in the fossil record, fully formed, and with no fossils yet discovered that form a transition from their parent groups." (Hickman 1)

The Hickman Authors also stated, "the fossil record suggests that ... it has seldom been possible to piece together ancestor-dependent sequences from the fossil record." (Hickman 2)

In 1993, Biologist, Professor Ridley, stated, "In the fossil record, stasis—evolutionary lineages that persist for long periods without change—is common." (Ridley 1)

As stated above, these scientists are all evolutionists, and their statements don't need additional comment; they speak for themselves. One of Darwin's dreams was that after many more fossils would be found in the future, they would eventually fill in the gaps and support his theory of evolution. But, based on the above facts and quotations, we can conclude that this particular dream of Darwin has not come true. The fossils do not support the theory of evolution.

There is much more information on fossils in *The Cambrian Explosion* (Starkey 1).

Conclusions
Based on the abundance of evidence cited above, which includes information on the fossils of the Cambrian-Explosion animals, and the fossils of the animals which have existed during the past 540,000,000 years, we must come to several conclusions:

(1) The fully-developed and extremely complex animals of the Cambrian Explosion appeared suddenly and instantaneously, and they did not evolve from any predecessor animals. This proves that the theory of evolution was not valid during the Cambrian period.

(2) Each of the animals that came into existence and which lived on our earth during the 540,000,000 year period after the Cambrian Explosion arrived suddenly, fully developed, and without any fossils that showed it to be the result of gradual multi-step evolution. For each animal, the fossils which should represent intermediary phases of the evolution of that animal are totally missing.

This Chapter constitutes Proof #4 that evolution which is capable of causing one species of animal to evolve into another species has never occurred on this earth. And this Proof is based on the facts which are the fossils.

CHAPTER 16. THE BASIC NATURE OF CHANCE

Chance and its Alternative

Since the theory of evolution is totally based on events that occur by chance, it should be very important for us to clearly understand the basic nature of chance. So let's now consider this in some detail.

There are two basic processes by which an objective might be achieved on this earth. These are: (1) processes which are based entirely on chance, and (2) processes which are based on intelligence and knowledge. With respect to the origins of animals, we are interested in whether they originated from the actions of the natural forces of the earth operating strictly on the basis of chance, or by the application of intelligence and knowledge.

We would like to have some guiding principle the application of which would enable us to determine which of these two processes was responsible for the design and construction of the animals. I have developed just such a principle, and I will now state that principle, and we will then discuss several examples for which the principle can be applied. This will help us understand the basic nature of chance.

The Basic Principle of Chance vs Intelligence and Knowledge

This principle states that if it is observed that some objective has been achieved, and if a statistical analysis reveals that there is essentially no possibility that the objective could have been achieved by pure chance alone, then it can be concluded that the objective was achieved by the alternative process, intelligence and knowledge We will now discuss several examples which illustrate the application of this principle.

Finding the Home Port on Lake Erie

I own a boat, which my wife and I use on Lake Erie. We frequently cross the Lake from Ohio to Canada, and, after visiting Canada we must cross the Lake in reverse to get back to our home port, which is on the Ohio shore just east of Port Clinton. Suppose on one voyage, when we were in the middle of Lake Erie heading for our home port, I decided that I would like to get back to the home port by chance. I would take my hands off of the steering wheel and let the boat go wherever it would, by chance. What is the statistical probability that I would ever get back to the home port?

52

The shores that surround Lake Erie are 600 miles in length. There are 5280 feet in each mile. The entrance to the channel of my home port is 50 feet wide. The boat, without my steering, is likely to go in any direction. Based on these facts we can calculate that the statistical probability that the boat, by chance, would find the entrance to the home port. It would be

$$\text{One chance in } (600)(5280)/(50) = 63,360.$$

If the boat came to shore at the wrong location, it would have to go out to sea again and try once more to find the home port. If we elected to stay on the boat until it finally found the home port by chance, we would probably die of old age before the boat ever found the entrance to our harbor.

On the other hand if we had applied our intelligence and our knowledge of navigating and piloting, we would, with 100% certainty, have found the home port at the end of the first try, and we would have gotten from the middle of Lake Erie to the Ohio shore in about two hours. This example shows the tremendous superiority of intelligence and knowledge over the process of chance.

Now suppose that there was someone on the shore at the home port who saw us come cruising in, and, to make conversation with that person, suppose I told him that I had taken my hands off of the steering wheel in the middle of the Lake, and the boat, by chance, and by steering itself, had found the entrance to our harbor. He would have called me a liar. Why? Because he understood the general principle that applies to an objective that might be achieved either by chance or by intelligence and knowledge.

Opening the Combination Lock on a Safe

I have in my office a safe that has in it one drawer that can only be opened by operating a combination lock. The lock has 100 numbers on its face, and, to open the lock, the arrow of the movable knob must be directed, in turn, to four different secret numbers. To open this lock the operator must rotate the knob clockwise four times and stop it at the 1st number. The knob must then be rotated three times counter-clockwise, and be stopped at the 2nd number. It must then be rotated clockwise two times and be stopped at the 3rd number. Finally, the knob must be rotated in the opposite direction and stopped at the 4th number.

To open this lock, the movable knob must be stopped at each of the four secret numbers. Also, the correct procedure must be used which involves the directions of rotation and the number of rotations that must be completed before stopping at the next number. However, to analyze this lock, we will forfeit the complexity involved in the process of applying the numbers, and we will perform a statistical analysis that will depend only upon employing each of the secret numbers in the proper order. Let's analyze the probability that this safe could be unlocked by pure chance.

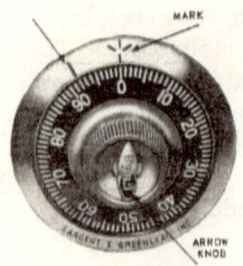

There would be one chance in 100 that the first number would be correctly applied, if its choice were based on chance alone. There would also be one chance in 100 that each of the other three numbers would be found by chance. Therefore the statistical probability that this safe could be opened by chance would be,

one chance in (100)(100)(100)(100) = 100,000,000.

This is one chance in 100 million! This means that there is essentially no possibility that this safe could be opened by the employment of pure chance alone. One chance in 100 million is essentially no chance at all.

On the other hand, this safe could be opened with 100% certainty in 5 minutes if the numbers were applied with the aid of intelligence and knowledge. This example, again, illustrates the almost infinite superiority of achieving an objective by intelligence and knowledge, rather than by chance alone.

Most people know of the great superiority of intelligence over chance, and they do not attempt to open a safe by chance. And most people also know that with intelligence and knowledge the safe can easily be opened.

Calling a Friend on the Telephone

Suppose you need to talk long-distance to one of your friends on the telephone, and suppose you do not know the friend's telephone number. You could simply operate the telephone on the basis of chance, and maybe you would stumble upon the right phone number. Let's calculate the statistical probability that you could, by chance, come upon the right number.

Each long-distance telephone number has 10 digits, and you would need to guess each one of these digits. There is one chance in 10 that you would correctly guess the first number, and another one chance in 10 that you would correctly guess each of the other 9 numbers. Hence the statistical probability that you could, by chance alone, come up with the correct telephone number would be,

One chance in (10)(10)(10)(10)(10)(10)(10)(10)(10)(10) = 10,000,000,000.

This is one chance in 10 billion! And this means that you could not come up with your friend's telephone number by applying the process of pure chance.

On the other hand, if you are intelligent and you have knowledge of your friends telephone number, and you know how to operate a telephone, then there is one chance in one that you could be successful in making your phone call. This again illustrates the almost infinite superiority of intelligence over chance.

Let's now apply our principle which relates to chance, to make sure that you clearly understand it and appreciate it. If you do succeed in calling your friend, then your friend knows that you somehow came up with the correct phone number and achieved the objective of calling him on the phone. And, if your friend learned that there was only one chance in 10 billion that you could have come up with the number by chance, your friend could then logically conclude that you learned of and used the number by applying your intelligence and your knowledge. The application of this principle of chance constitutes a proof!

A Ring of Rocks for a Campfire on a Mountain Top

I have spent many hours hiking and Jeeping in the Rocky Mountains. While on one such excursion, after hiking up a rounded hill, I came to a flat meadow on the top of the hill. There I came upon a group of rocks arranged in a circle, similar to the ring of rocks shown in the Figure.

RING OF ROCKS

Each rock was about 9" to 10" in diameter, there were 10 rocks in the ring, and the diameter of the ring of rocks was about 30". Other rocks of various sizes were scattered about in a typical random fashion.

I wondered how this circle of rocks came into existence in this wilderness area. This fire-ring could have come into existence by the random chance actions of natural forces of the earth. Possibly a group of rocks, all of about 9.4" in diameter, somehow, by chance, became selected and were transported from the surrounding area to a central location, and they were then arranged into a circle, all by natural forces. Such natural forces might include wind, rain, hail, snow, flooding, freezing, thawing, earthquake, lightning, ice movement, cosmic rays, radiation from the sun, etc., all following the laws of physics, chemistry, and mathematics.

Or, alternatively, they could have come into existence by the actions of intelligent design. Possibly some human being selected a certain size rock, carried some rocks of that size to the location of the ring, and then arranged them into the form of a circular ring to construct a fire-ring in which he could build a camp-fire.

To determine which of these two processes might have designed and constructed the fire-ring I made some reasonable geologic and engineering assumptions concerning the movements of rocks by natural forces, and I calculated the statistical probability that the fire-ring could have come into existence by natural forces alone. I assumed that the rocks were selected from within a circle on the mountain having a radius of 50 feet. A much more detailed analysis of the origin of this fire ring is presented in *The Cambrian Explosion*, (Starkey 6). Using the data which I arrived at using my reasonable engineering assumptions, I was able to calculate the probability that the fire-ring could have been designed and constructed by the natural forces of the earth. The answer turned out to be:

one chance in   8,320,000,000,000,000,000,000,000,000,000.

This is essentially the equivalent of concluding that there is no possibility whatever that natural forces of the earth, alone, acting on the basis of chance, could have designed and constructed this simple fire-ring in the wilderness. On the other hand, if the fire-ring were designed and constructed by the application of intelligence and knowledge, the fire-ring could have been completed, with 100% certainty, in less than one hour. Applying the principle of chance we can conclude that intelligence and knowledge were responsible for creating this fire-ring.

The Principle of Chance Applied to the Theory of Evolution

We can apply the basic principle of chance to the theory of evolution. In Chapter 12 of this book we proved that for one animal to evolve into another the natural forces of the earth would need to find and act upon the gamete cell that was destined to produce the next offspring of a predecessor animal. For our typical animal there would be only one chance in 100 billion that this cell could be found by chance. Then, in Chapter 13 of this book we proved that there would be only one chance in 1.3 trillion that just 15 DNA atoms could be rearranged to produce some small beneficial change in the predecessor animal. Even though the order of magnitude of the number of atoms that would need to be rearranged to produce any noticeable change in the predecessor atoms would be millions or billions, we are going to ignore this fact and, to provide an ultraconservative analysis, we will combine the two probabilities that are defined above, namely one chance in 100 billion to just find the gamete, and one chance in 1.3 trillion to rearrange some DNA atoms. Just combining these two factors would produce a statistical probability of:

$$\text{one chance in } 130,000,000,000,000,000,000,000.$$

This is the probability that the natural forces of the earth could find the gamete cell in a predecessor animal and correctly rearrange just 15 atoms in its DNA molecule.

We can now apply the basic principle of chance that was cited at the beginning of this Chapter. It asserts that if an analysis, such as that given above, proves that there is zero statistical probability that a certain objective could be accomplished by chance, and if we observe that that objective has in fact been accomplished, then we can conclude that that objective must have been accomplished by the application of intelligence and knowledge. We can observe that animals have been designed and constructed. Therefore, based on the application of our principle of chance, we can conclude that these animals must have been designed and constructed by the application of intelligence and knowledge, and not by the theory of evolution.

CHAPTER 17. ALLEGED SCIENTIFIC EVIDENCES FOR EVOLUTION

Even after learning of the statistical impossibility that the DNA atoms of one animal might be rearranged by chance, to produce a new animal, and even after learning of the reasons why scientists accept the theory of evolution, some readers still ask the question, "But isn't it true that scientific evidences support the theory of evolution"?

Ernst Mayr announced in the July 2000 issue of *Scientific American*, "No educated person any longer questions the validity of the so-called theory of evolution, which we now know to be a simple fact." If asked to explain how Darwin's theses have been confirmed, most college professors, high-school teachers, and educated people of the media or the general public will cite each of a group of standard evidences that are commonly pointed to by evolutionists. Most of these people learned of these standard evidences from textbooks and biology courses taken in school. These evidences include the following: Darwin's tree of life, similar embryos, the reptile/bird Archaeopteryx, horse evolution, four-winged fruit flies, similar bone designs, The Miller-Urey experiment which suggests how life could have begun, evolution by the natural selection of moths, and the evolution of some of the finches on the Galapagos Islands. To many biologists these are the scientific evidences that prove that evolution is a fact.

In the Chapters that follow we will briefly discuss each of these evidences and we will let you judge the validity of these standard citations. As we discuss these examples we will apply, wherever it is applicable, the central concept of this book, namely the statistical impossibility of rearranging the atoms of a DNA molecule such that one species might evolve into another.

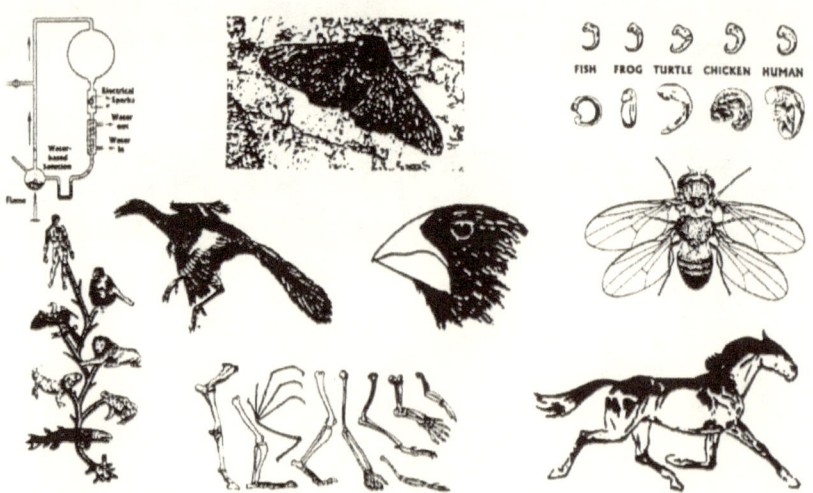

CHAPTER 18. THE FAMILY TREE OF ANIMALS

Animal History Suggests a Tree

Family Tree

It has been estimated that there have existed about 50,000,000 species of animals, ranging from the superbly designed human being to the small but remarkably complex single-cell animal. These species include protozoa, trilobites, sponges, squid, spiders, insects, fishes, amphibians, reptiles, birds, and mammals. Paleontologists and geologists tell us that these animals appeared at various times during the past half-billion years. Evolutionists assume that every animal evolved from some predecessor animal, and that all animals evolved from a first single-cell creature. Based on these assumptions, and the observation that different animals appeared at different times, evolutionists have suggested that the animal kingdom must be similar to a tree, of which the first single-cell animal was the root, and the subsequent animals could be likened unto fruit on the branches of a tree. Then they have attempted to place the various animals at various locations on the tree. Figure 18.1 shows Darwin's tree of life, and Figure 18.2 shows a tree of animals which is similar to Darwin's rendition. But in Figure 18.2 there are circles each of which represents an animal, and the lines between circles represent lineages. However, as appealing as may be the tree-of-life depiction to the evolutionist, objective scientific facts and analyses reveal that this concept is flawed and it does not support the theory of evolution. Consider the following facts.

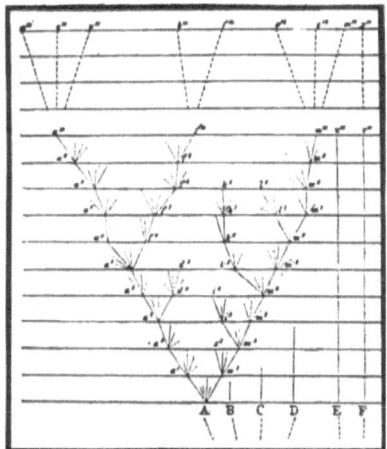

Figure 18.1 Darwin's Tree of Life

The Cambrian Explosion Turns the Tree of Animals Up-side-down

First, the Cambrian Explosion turns the tree up-side-down. Contrary to the notion that the tree of animals started with a root which consisted of one single-cell animal, the facts provided by the fossils reveal that animal life essentially began by the sudden appearance of dozens of complex, fully-developed animals. These animals came into existence about 543,000,000 years ago and they all appeared within a period of time which was from 5 to 10 million years long. This, geologically, is an instant of time, amounting to only about one tenth of one percent of the age of the earth at that time. This sudden appearance of animals is widely referred to as the Cambrian

Explosion. See *The Cambrian Explosion* (Starkey 10). And, these animals contained representatives from most of the phyla of animals that exist today. During the Cambrian explosion, of the 18 major phyla of today's animals, 16 were represented. Only the worms of Nematoda and Platyhelminthes were missing, and they appeared slightly later. Thus, the Cambrian explosion turned the animal tree of life up-side-down and eliminated it as a tree. It produced a thicket, not a tree.

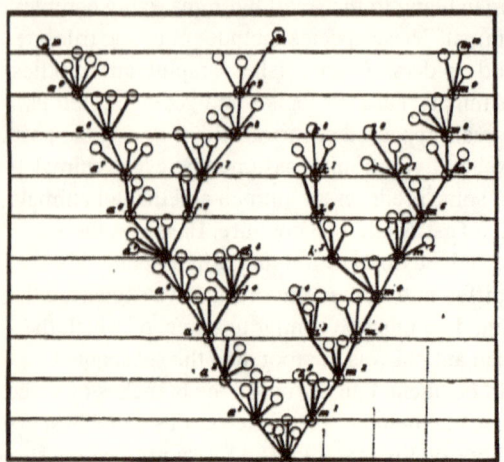

Figure 18.2 A Tree of Life Similar to Darwin's

Secondly, please refer again to Figure 18.2. This Figure shows the first single-cell animal at the bottom, which might be the root, and then the lines are drawn above the root, which are the branches, and finally, its circles represent the fruit, which are the animals. Above the first single-cell animal, each circle at a node or a tip represents a new species which has evolved from the lower species which is near it and which is closer to the root.

Tree Branches Represent Evolutionary Lineages

The information contained in Figure 18.2 reveals another fatal flaw in the evolutionists' tree of animals which eliminates it as a possibly accurate representation of the origin of the species. Please direct your attention particularly to the lines on Figure 18.2 which connect the circles. What does each of these lines symbolize? It represents the *evolution* of one species into another. It doesn't just indicate that the species of a higher circle on the drawing at the end of a line came into existence at a later time than the lower circle, it is intended to show that the upper circle evolved from the lower circle by descent-from-an-ancestor evolution. But, if it could be shown that evolution, itself, cannot, does not, and never did occur, then all of the lines on Figure 18.2 should be removed. In Chapter 13 we showed that the only possible way for one animal to evolve into another would be for millions of atoms to be moved, by chance, to millions of other specific locations in the DNA molecule of the newer species. But we proved that there would be only one chance in 1.3 trillion that just 15 atoms could be so moved, and there is no chance at all that millions could be moved. Also, whatever mutation-causing agent which might be thought to be responsible for the atom relocations would have to find the gamete cell of the ancestor, which would require it to find one cell out of billions, by chance. Since there is no evolutionary process which can beat these odds, we have to conclude that evolution did not take place, and all of the lines which represent branches on Figure 18.2 should be eliminated.

The Lines of Figure 18.2 should be Removed

What then should the tree of animals look like? The first single-cell animal of Figure 18.2 should be replaced by a horizontal row of circles which would represent the many animals of the Cambrian explosion. Then, each of the circles above the bottom row should represent the sudden appearance of a new species at some time during the past 543,000,000 years. Each new species was the product of the design-and-construction efforts of an intelligent designer, who was capable of collecting the required atoms of H, O, N, C and P and placing them in the correct positions in the DNA molecule of every cell of the new species. These animals did not evolve from some earlier animal. They were all constructed from scratch. And, finally, on the top row of Figure 18.3, we see the animals of today, including fish, reptiles, birds, mammals, and man. The general conclusion to be reached, based on these assertions, is that all of the lines of Figure 18.2 should be eliminated, and a bottom row should be inserted instead of the single-cell circle. These changes would produce a substitute depiction which might more accurately represent the origin of species. This would look like Figure 18.3. In this Figure there are no lines connecting any circle with any other circle. And rather than displaying a tree, the resulting picture looks more like a thicket. The dotted lines simply represent the Cambrian Explosion and today.

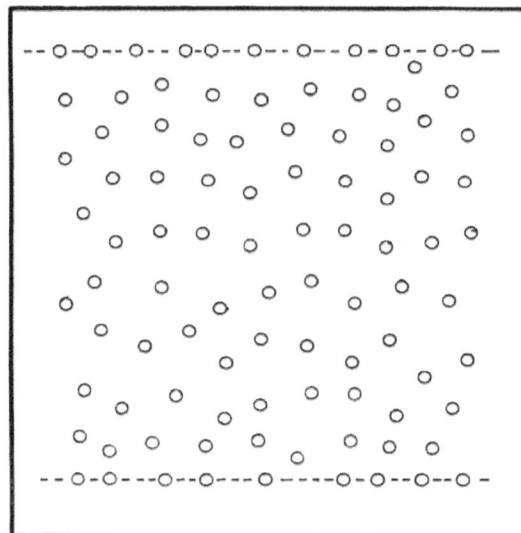

Figure 18.3 An Accurate Representation of the Origin of the Species

Scientists Agree that Figure 18.3 represents what the Fossils reveal.

The assertions that the tree-figure of Figure 18.2 is scientifically indefensible, and that the depiction of Figure 18.3 is the correct one, are further supported by the findings and statements of research scientists. In Chapter 15 we recorded about a dozen statements of recognized scientists which asserted that the branches of a tree do not exist. Dr. Patterson's statement is typical and is worth repeating here.

Dr. Colin Patterson, senior paleontologist at the British Museum of Natural History, wrote to Luther Sunderland, "You say that I should at least show a photo of the fossil from which each type of organism was derived. I will lay it on the line—there is not one such fossil for which one could make a watertight argument." (Patterson 1)

Evolutionists Suggest that Molecular Phylogeny offers a New Tree

In recent years scientists have learned more about DNA, and this has introduced a somewhat different tree-of-life concept which is called molecular phylogeny. The word "phylogeny" is defined as "a taxonomic division of the animal kingdom." The word "taxonomy" is defined as "a classification of organisms." The suffix "geny" is defined as "a manner of origin or development." By putting these all together we may conclude that "phylogeny" is the manner of the origin of the phyla of animals. The word "molecular", as it is applied here, refers to the DNA molecule, one of which is in every cell of every animal. But this molecule is different and unique for each individual animal. Molecular phylogeny then refers to the attempt by biological scientists to study segments of the DNA molecules of animals, note similarities, and then speculate concerning the lineages which may be deduced from these similarities to produce a tree of life, based on the DNA molecules.

Using molecular phylogeny, scientists also attempt, based on assumed frequencies of mutations, to estimate the number of years that must have elapsed from the time that a common ancestor split, to the present.

Is Molecular Phylogeny Flawed?

There are many faults, defects and shortcomings of this relatively new science called molecular phylogeny. Consider the following:

1. Since fossils do not have any DNA molecules, only the animals that exist today can be involved in these analyses. The fossils contribute nothing to molecular phylogeny.

2. It is assumed that genes, which are specific arrays of atoms in a DNA molecule, determine evolutionary relationships. This is nothing but another speculative assumption.

3. Frequencies of mutations are estimated, based on speculative assumptions which are inferred from other considerations, and these frequencies are highly hypothetical.

4. The entire science of molecular phylogeny is totally based on the assumption that every animal evolved from some predecessor animal. Evolution is merely *assumed* to exist. Not only have we shown this to be a false assumption, but also it would then be circular reasoning to assert that molecular phylogeny supports the theory of evolution.

5. The results obtained from these analyses when performed by various investigators vary wildly.

It might be said that molecular phylogeny is more a game of sophisticated puzzle-solving than a respectable science.

Finally, to be sure that you appreciate the most important concept in this book, please consider, again, the following facts. Admittedly these facts are repetitions, but the concept is so significant that it bears repetition.

1. A species of animal is absolutely determined by its array of atoms in its DNA molecule.

2. For an animal of one species to evolve into a different species it is necessary for millions of atoms to be moved from the DNA array of the first animal to the DNA array of the second.

3. To accomplish this movement, first, some mutation-causing agent must find the gamete for the next offspring of the first animal, which, statistically, is an impossible task.

4. Secondly, if the gamete were found, there is only one chance in 1.3 trillion that just 15 atoms could be successfully moved, and there is no chance that millions could be moved.

Based on these mathematical facts, and on the facts provided by the fossils, we can conclude that evolution has never taken place, and the whole concept of an animal tree of life is legally moot and without significance. The tree does not exist.

As inconceivable as it may seem, it is a fact that the animal tree of life is still presented in most college and high-school textbooks of today, and it is taught to the students as a scientific evidence in support of the theory of evolution.

CHAPTER 19. IDENTICAL EARLY EMBRYOS

Identical Embryos Suggest Descent from a Common Ancestor

Darwin, and other evolutionists in the mid 1800's, observed that each species of animal began as a single cell, then developed through several embryonic stages and ultimately became an adult. It then occurred to them that a very similar series of events may have taken place as the animals of the earth evolved from the first single-cell animal, then developed through the evolution of several primitive animals, and finally evolved to the complex mammals and man. Based on this theoretical observation they deduced that there might be a connection between embryonic development and evolution. They then further suggested that if it could be shown that, at a particular stage of development, all of the embryos of a wide variety of animals looked alike, they might conclude that this would provide evidence that all of these different animals were evolved from a common ancestor.

Haeckel's Drawing of Identical Embryos

Among the early evolutionists who thought along these lines was the German biologist, Ernst Haeckel (1834-1919). Based on his studies of embryos and his interpretations of these studies, Haeckel prepared some drawings of the embryos of several classes of vertebrates including fish, salamander, tortoise, chick, hog, calf, rabbit and human (Haeckel 1). Haeckel's drawings are reproduced in Figure 19.1.

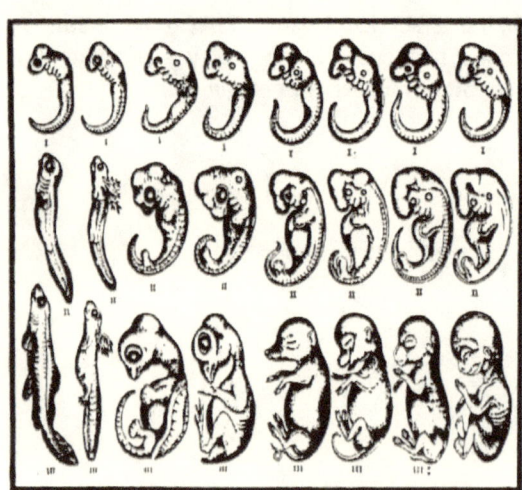

Figure 19.1 Haeckel's Drawing of Identical Early-stage Embryos (See Top Row)

Note that all of the embryos in the top row, which correspond to an early stage, are almost identical. To Haeckel and Darwin this represented the common ancestor in the analogous process of evolution. Note, also, that by the time that the embryos are as depicted in the bottom row, each embryo is clearly different from the others. This represents the various different animals that have allegedly evolved from the common ancestor.

Darwin, of course, as an evolutionist, enthusiastically welcomed the publication of Haeckel's drawings, and, in concert with Haeckel's thoughts, Darwin wrote, "Generally the embryos of the most distinct species belonging to the same class are closely similar, but become, when fully developed, widely dissimilar." And

Darwin added that since humans and other vertebrates "pass through the same early stages of development, ... we ought frankly to admit their community of descent." (Darwin 2)

Haeckel's Drawings were Fakes

However, as was discovered by embryologists of Haeckel's day, and which has been confirmed by numerous embryologists from then until the present day, Haeckel's drawings were faked. They are a fraudulent deception. Actually, early embryos are not identical or even nearly identical, but rather they are clearly distinct, and they relate to their adult physiologies. But Haeckel drew them as being identical. Today, all embryologists agree that Haeckel's drawings were a fraud.

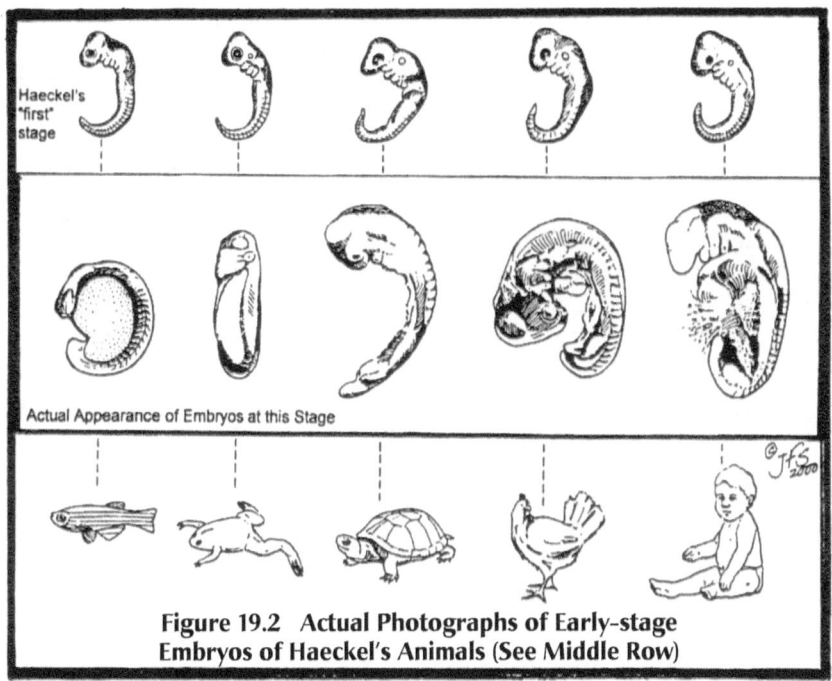

Figure 19.2 Actual Photographs of Early-stage Embryos of Haeckel's Animals (See Middle Row)

Copyright Jody F. Sjogren 2000. Used with permission.

Beginning in the day of Haeckel and Darwin, and throughout the past 150 years, many scientists have published repudiations of Haeckel's drawings or their interpretations. These include the following scientists: von Baer in 1835, Sedgwick in 1894, Lillie in 1919, Garstang in 1922, Gould in 1925, de Beer in 1958, Ballard in 1976, Oppenheimer in 1987, and Elinson in 1987. Statements made by most of the above-listed scientists are more fully explored by Jonathan Wells in his book, *Icons of Evolution* (Wells 2). Haeckel's contemporaries repeatedly criticized him for his drawings, and he was frequently accused of fraud during his lifetime

Actual Photographs

In 1997 British embryologist Michael Richardson and his associates took actual photographs of embryos which correspond to Haeckel's top row (Richardson 1). They are shown in Figure 19.2 The actual photographs are shown in the middle row, and Haeckel's drawings are shown in the top row. The obvious conclusion must be that Haeckel's drawings misrepresented the truth.

The Word "Descent" should be Removed from our Lexicon

Let's now add to the above discussions on embryos, the new facts that have been presented in this book. Based on the facts presented in this book, the entire matter of embryos, Darwin's thoughts and Haeckel's drawings are moot and of no consequence. Why? Because we have shown that for one animal to evolve into another absolutely requires the addition or rearrangement of millions of atoms in the DNA molecule of the next-to-be-used gamete, and there is only one chance in 1.3 trillion that just 15 atoms could be successfully moved by chance. Hence there is no chance that millions could be moved. In the minds of evolutionists, the embryos represent examples of the concepts of "descent from a common ancestor" and "descent with modification."

But if the required number of atoms cannot be moved by chance, then, in the phrase "descent from a common ancestor" or in the phrase, "descent with modification" the word, "descent" should be eliminated and then the phrases have no meanings. The word, "descent" means evolved, and we have proved by mathematical analyses that evolution could not, and has not, ever occurred. Haeckel's embryos should be an embarrassment to scientists, and the alleged identical embryos do not support the theory of evolution.

But Haeckel's drawings, and the assertion that they constitute scientific evidence for the theory of evolution, still appear in many college and high-school textbooks which are in use today. This reveals the zeal with which atheistic evolutionists promote their religion, even to the extent of reprinting what is commonly known to be a fraud.

CHAPTER 20. ARCHAEOPTERYX, THE MISSING LINK

The Archaeopteryx Fossils Appeared to be a Missing Link

A fossil which appeared to be the missing link between reptiles and birds was discovered in Germany in 1861. This fossil is now in London. Another similar fossil was found in 1877 and it is now in a museum in Berlin. Subsequently, six more similar fossils have been found. All eight came from a limestone quarry in Solnhofen, Germany. Herman von Meyer named these fossils Archaeopteryx, which means "ancient wing." These fossils had wings and feathers, which made them bird-like, but they also had teeth, long tails and claws on their wings, which made them reptile-like. These fossils, especially the Berlin fossil, have been so well preserved by the fine-grained limestone in which they had become buried that the bird-like feathers, and the reptile-like jaws, teeth and tails could all be carefully studied. The discovery of the Archaeopteryx fossils understandably delighted the evolutionists of Darwin's time, evolutionists of the past 140 years, and teachers and textbook writers of today. They all point with pride to the missing link which shows that birds evolved from reptiles, and Archaeopteryx, which has recently been identified as a bird, is widely cited as an important piece of scientific evidence in support of the theory of evolution.

Missing Link Theory has Problems

But a more careful, and a more scientific study of the matter reveals that these evolutionists have at least five major problems with their interpretations of the significance of the Archaeopteryx fossils.

1. The evolutionists must ignore the fact that the mathematical analyses of this book prove that evolution never occurred.

2. There are no intermediate fossils which link Archaeopteryx to modern birds, and most informed evolutionists of today have concluded that our birds did not evolve from Archaeopteryx.

3. There are no fossils which might link Archaeopteryx to the ancestors from which it might have evolved, such as some reptile.

4. All of the ancestors and descendants of Archaeopteryx seem to have become extinct, and Archaeopteryx seems to have come into the world from nowhere, and then it became extinct.

5. The morphological, or anatomical, differences between birds, which can fly, and reptiles, which crawl on the ground, are so different that it is inconceivable that even zealous evolutionists, who emphasize anatomical similarities, could accept Archaeopteryx as being a missing link.

Problem number 1 is discussed elsewhere in this book, and it need not be repeated here. The problems of 2, 3 and 4 are self-explanatory. But problem 5 deserves to be explored in more detail.

Birds and Reptiles are Extremely Dissimilar Morphologically

We are attempting to determine the probability that a reptile might have evolved into a bird. One of the most important factors in making this determination is the degree to which these two animals are anatomically similar. It happens that these particular two animals are extremely different morphologically. Obviously if two animals are very dissimilar, this fact makes it very improbable that one could be said to have evolved from the other. Since the complexity of a bird is a factor in our study, I thought it might be informative for us to study in some detail the design characteristics of a bird, and compare them with the design characteristics of a reptile.

A Dead Bird Taught me the Secret of its Wings

For many years I have marveled at the mechanical capabilities of birds. As I was studying birds, I observed that birds can fly directly upward by flapping their wings up and down. Based on the readily observable characteristics of a bird's wings, and my engineering analyses of how these wings could impart lift to a bird, I could not explain how this lift could be developed. When the wings move down the bird is lifted, but when the wings move upward, an equal downward force is developed. Since these cancel, I could not explain how a bird could fly straight upward. So I inquired from many experts, including biologists, aeronautical engineers, authors of books on this subject, and others. None of them could explain how a bird can fly straight upward merely be flapping its wings.

Figure 20.1 Bird Which Fell out of Tree Right in Front of Me. 7.5-inch Wing-span.

Finally, I discovered the answer! In fact a bird showed me the answer. As I was walking along the gravel road on my home property, a bird fell out of a tree right in front of me. It was dead. But it was still warm and very flexible. I picked it up and examined its wings. There I discovered what I believe is one of the most clever and remarkable designs in all of the animal kingdom. And, apparently, few other engineers or scientists have discovered what this bird taught me. I put the bird on the scanner of my computer and took its picture. Figure 20.1 is the picture of this bird, which miraculously taught me a lesson in engineering.

The feathers of the wings are designed in such a way that when the air pressure is upward under the wing, the feathers lay flat against each other and a continuous and relatively air-tight flat surface is presented to the pressure, so that a large upward lift can result from a downward motion of the wing. Then, when the wing is moved upward,

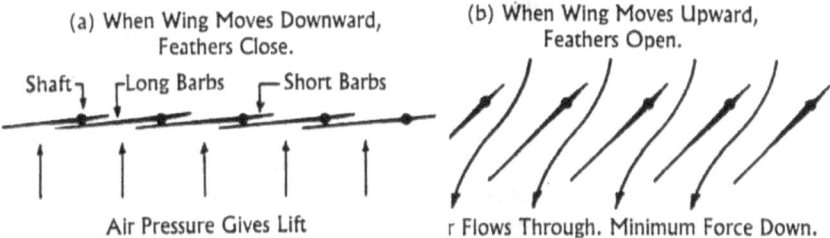

Figure 20.2   How a Bird's Flapping Wing gets Lift

each of the feathers rotates, like a Venetian blind, which produces a large gap between adjacent feathers, through which air can move without much resistance. As the wing is moved upward, air flows freely through these gaps and very little downward force is generated. I could blow on the upper side of the wing of this bird and the feathers would all open up the gaps. Blowing on the lower side closed the gaps and I could feel the greater force caused by the blowing The wings acted as one-way valves. In order for the feathers of a wing to act as a one-way valve it is necessary for the quill and shaft of each feather to be largely off-center, as shown in Figure 20.2. When you see a bird feather, look to see if the shaft is off-center.

Birds have Many Special Features that Enable them to Fly
In order for birds to fly they not only must have uniquely designed wings and wing-powering mechanisms, but they must be light in weight. Their design is replete with many unique features, all of which serve the purpose of producing a flying machine that is light in weight. The bones are hollow and thin-walled, but they are reinforced with internal braces that prevent mechanical failure by elastic instability. They digest their food rapidly and excrete it quickly, to lighten their digestive systems. Their lungs are supplemented with air sacs to provide the high-capacity respiratory system needed for flight. Their hearts have very high power-to-weight ratios, achieved by having high operating speeds. Any machine that operates at high speed can produce more power for a given amount of weight than a slower-speed machine. The heart of a black-capped chickadee beats at 500 times per minute when asleep, and 1000 beats per minute while in flight.

Birds do not carry their young in their bodies like land mammals do, nor do they carry large numbers of eggs in their bodies as do fishes and reptiles. Birds generate one egg at a time and lay it as soon as it is produced. Birds also have a unique system for perching on branches which utilizes the weight of the bird's body to pull on certain tendons which, in turn, force the bird's feet to clasp the branch tightly without the need for any muscles to pull on these tendons. A bird can go to sleep and this clever clasping mechanism keeps it firmly attached to its perch. It uses the force of gravity rather than the force of muscles. There is a much more detailed treatise on birds in *The Cambrian Explosion* (Starkey 11).

Birds and Reptiles are Totally Different Anatomically

As an engineer and a machine designer I have studied these features of a bird, and I can testify with engineering certainty that these clever devices are too complex and well designed to have come about by chance. They must have been designed by an intelligent designer. All of these almost unbelievable engineering design characteristics of a bird are in stark contrast with the characteristics of a reptile. The reptile is too heavy to fly and it does not possess any of the many special features mentioned above which enable the bird to fly.

Thus, we can conclude, based on their vastly different anatomical differences, that birds did not evolve from reptiles. Also, based on morphologies and on the fossil records, birds did not evolve from Archaeopteryx and Archaeopteryx did not evolve from a reptile. Apparently, Archaeopteryx appeared suddenly, had no common ancestor with any other creature, and then quickly became extinct.

Archaeopteryx was a Fascinating and Unique Animal

How much faith in the atheistic religion does it take to believe in something that is scientifically improbable and mathematically impossible? It takes less faith to believe that some very creative intelligent designer, possibly in a playful mood, designed, placed the atoms in the right locations in the DNA molecules, and constructed the very unique animal, Archaeopteryx. It must have been great fun. But the discovery of Archaeopteryx does not constitute any scientific evidence in support of the theory of evolution.

Nevertheless, it is a fact that Archaeopteryx is still presented in most college and high-school textbooks of today as the missing link between a reptile and a bird, and it is taught to the students as a scientific evidence in support of the theory of evolution.

CHAPTER 21. THE EVOLUTION OF THE HORSE

Horse-like Fossils Suggested Evolution

There are 100,000,000 fossils in the various museums of the world. These fossils are of animals that have existed on earth during the past 543,000,000 years, since the Cambrian explosion. After the dinosaurs became extinct, about 65 million years ago, the mammals came into existence, and among these mammals several fossils have been found which are anatomically similar to our horses of today. Biological scientists have studied these horse-like fossils, organized them into species, and assigned a name to each species.

Heads	Fore Foot	Hind Foot
Equus	One Toe	One Toe
Protohippus	Three Toes	Three Toes
Mesohippus	Three Toes	Three Toes
Proterohippus	Four Toes	
Hyracotherium	Four Toes	Three Toes

Figure 21.1 Early 1900's Concept of the Evolution of the Horse

In the early 1900's evolutionists studied several of these species, determined their ages, and selected which, according to the evolutionists, showed that our modern horse has evolved from these prior animals. Figure 21.1 shows such a group, and their anatomical characteristics are noted. A small group of fossils from western United States seemed to indicate that, as the horses evolved, during the past 50 million years, they got larger in size and their numbers of toes diminished from four to three, and then to one. The message that Figure 21.1 is intended to convey is that the small, four-toed animal, Hyracotherium, of 50 million years ago, evolved to produce, in succession, Protorohippus, Mesohippus, Protohippus, and, finally, the modern horse, Equus. For many years this sequence of horse-like animals was touted in museums and textbooks as scientific evidence in support of the theory of evolution.

But later in the 1900's it became apparent that many other species of horse-like animals had also existed, and the alleged lineage of the horse was not a clear straight-line sequence, but consisted more of a multi-branch tree or thicket. Therefore the evolutionists were forced to add more animals to the alleged lineage, and pictures such as that shown in Figure 21.2 were composed. However, within the thicket of animals, to show the lineage from Hyracotherium to Equus, it was necessary to pick some group of fossils through which a heavy black line could be drawn to represent the direct line of descent between the two. Note this line in Figure 21.2. Also observe that along the heavy line of Figure 21.2 there are only two of the four fossils which are shown as the lineage to Equus in Figure 21.1.

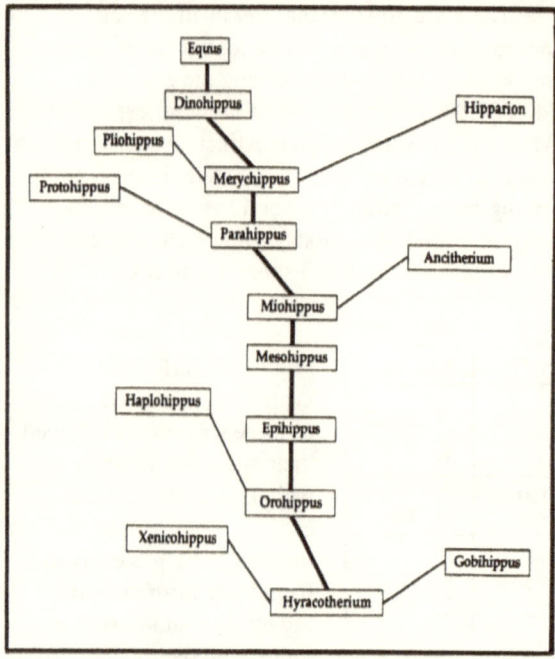

Figure 21.2   Later 1900's Concept of the Evolution of the Horse.

Copyright Jody F. Sjogren 2000. Used with permission.

The message that is intended to be conveyed by Figure 21.2 is that even though there have been many horse-like animals whose fossils have been found, there still was a group of them which constituted the evolutionary lineage from Hyracotherium, of 50 million years ago, to Equus, our horse of today. Let's examine this alleged lineage.

In Figure 21.2 there are sixteen boxes each of which contains the name of a species of horse-like animal for which a fossil has been found. The fossils associated with these species constitute real scientific information.

The Lines of Figure 21.2 Represent the Assumption of Evolution

But on Figure 21.2 there are many *lines* drawn which connect these boxes. What information are these lines intended to convey? Each line has been drawn to represent the evolution of the species in one box to the species in another box. But do these lines represent proven scientific facts? The answer is, no. Each line is nothing but an assumption. Are there any in-between fossils which show how one species gradually changed into another? Are there any fossils having 2-1/2 toes or 3-1/2 toes? Are there any fossils which contain new features that are half finished? The answers are all no. Each of these lines, alone, *assumes* that evolution took place, and that it involved the two species the boxes for which are connected together by a line. We will show below that these lines should be eliminated, and the term, "evolutionary lineage" should be deleted from your lexicon of scientific words.

With the Lines Removed Intelligent Design Replaces Evolution

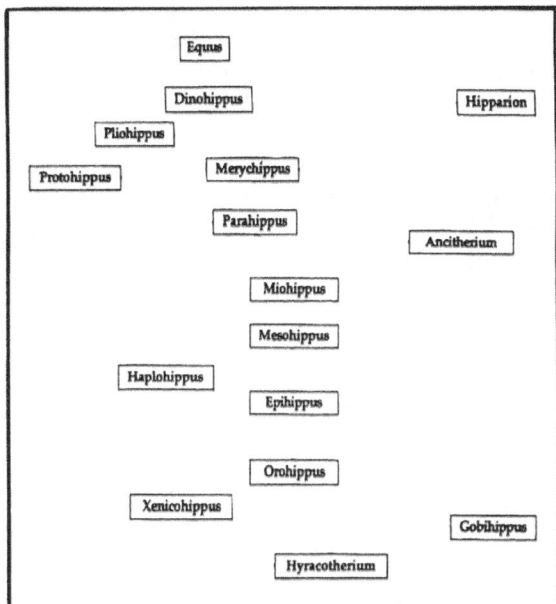

Figure 21.3 An Accurate Depiction of the Origins of Horse-like Animals which is Consistent with the Facts Provided by the Fossils.

Copyright Jody F. Sjogren 2000. Used with permission.

Figure 21.3 has been prepared and it lists all of the species which are identified in Figure 21.2, but the lines have been removed. Let's examine the message of Figure 21.3. This Figure represents a group of fossils of species each one of which was created by an intelligent designer at a particular time in history. Each of these species, except Equus, was designed and created independently of all of the other species, and it eventually became extinct. Equus is still with us. Now we should ask the question as to which depiction, Figure 21.2 or Figure 21.3, accurately represents the historical truth.

Let's recall again that there is no way that evolution could take place other than by adding to or rearranging the atoms in the DNA molecule of the gamete cell of the predecessor animal. In Chapter 12 we learned that the statistical probability that a mutation-causing agent could find the gamete cell of a typical small animal is about one chance in 300 billion. In Chapter 13 we learned that the probability that an agent could successfully move just 15 atoms is one chance in 1.3 trillion. Therefore evolution cannot occur, and it never has occurred. Hence, in the case of the horse-like animals for which fossils have been found, the evolutionary process which is depicted by the lines of figure 21.2 did not take place. But, rather, since we have proved that evolution has never taken place, the sudden appearance of these animals due to intelligent design, as depicted in Figure 21.3, must be the correct interpretation of the fossil data.

For all Animals the Concept of Lines of Descent is Fallacious

Actually, the alleged evolution of the horse is no more convincing than the evolutionary lineages which are presented in the Table which is shown in Chapter 22 of *The Cambrian Explosion* (Starkey X), which covers all of the members of the animal kingdom. That Table describes lineages in words rather than by such lines as those of Figure 21.2. But those words are so ambiguous, indefinite and weasel-like that the lineages which they are supposed to represent should clearly be treated as assumptions, certainly not facts. Similarly, the lines of Figure 21.2 should be recognized as pure assumptions, and, based on our mathematical analyses, they should be eliminated. And such depictions as Figure 21.2 and the alleged evolution of the horse which it represents do not constitute scientific evidence for the theory of evolution.

But nevertheless, it is a fact that the evolution of the horse as depicted in Figure 21.2 is still presented in most college and high-school textbooks of today as the evolutionary lineage of our modern horse, and it is taught to the students as a scientific evidence in support of the theory of evolution.

CHAPTER 22. THE FOUR-WINGED FRUIT FLIES.

The Normal Fruit Fly has Two wings and Two Balancers

The wild fruit fly, Drosophila melanogaster, normally has a middle region, or thorax, upon which are mounted its two wings. Just behind the wings are mounted a pair of weighted cantilever beams, called, balancers. In my engineering opinion, these balancers have been designed so that they vibrate up and down oppositely to the wings and they counterbalance the vibratory forces of the wings and hence provide smooth, rather than jerky, forces on the thorax when the insect flies. Electric hair clippers have in them just such a vibrating cantilever beam to counteract the vibrations of the clipper mechanism as the clipper is held in the hand of the barber.

The two wings and balancers can be seen in Figure 22.1.

Figure 22.1 A Normal Fruit Fly which has Two Wings and Two Balancers

Mutations and Breeding Produce a Fruit Fly with Four Wings

It is commonly known that mutations in animals do occur in nature. In the mid 1900's several biological researchers, including Cal Tech geneticist, Ed Lewis (Lewis 1) noticed that some fruit flies did experience various mutations which affected the thorax and the balancers. After much experimentation which involved breeding these mutants the

Copyright Jody F. Sjogren 2000.
Used with permission.

researchers finally produced a fruit fly in which the balancers apparently turned into a second pair of wings. Such a fruit fly is shown in Figure 22.2. The researchers had produced a four-winged fruit fly. Again, evolutionists greeted this accomplishment with great enthusiasm, and in 1995 Ed Lewis was given a Nobel Prize for his work. In 1999 Peter Raven and George Johnson (Raven and Johnson 1) wrote, "Genetic change through mutation and recombination provides the raw materials for evolution."

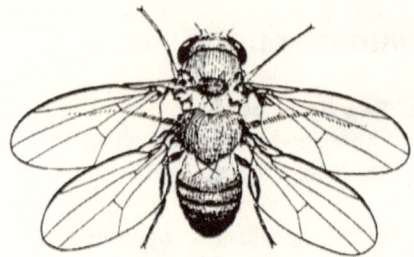

Figure 22.2 A Fruit Fly which has been Modified by Mutations and Breeding. It has Four Wings and no Balancers

Copyright Jody F. Sjogren 2000. Used with permission.

The Four-winged Fly is Inferior

We might now ask several questions. First, is the four-winged fruit fly better than the two-winged fly? Could it be described as an evolutionary advancement? The answer to these questions is, no. The second pair of wings do not have any muscles. They are not functional, and in fact, they do nothing but hinder flight. Also, the balancers are missing, having been turned into useless wings, and their useful counterbalancing function is gone. Also, the four-winged flies have difficulty mating. If these fruit flies were to be released into the wild, where they would be acted upon by natural selection, they would quickly die and their kind would become extinct.

Secondly, is it likely that such controlled mutations and breeding which produced the four-winged fly would ever occur in the wild? The answer is no. To produce this special fruit fly it was necessary to find three particular different mutants and then breed them by a special procedure. The statistical probability that this would occur in the wild is essentially zero.

Thirdly, of all of the millions of beneficial mutations that evolutionists believe must be occurring in the 50,000,000 species of animals in existence today, how many really significant examples of beneficial mutations have been found. The answer is zero. And the four-winged fruit fly is not an exception. It has not experienced a beneficial mutation.

Mutations Change Animals for the Worse

All that the above-described research does show is that mutation-causing agents can change an animal, but in the fruit-fly research the offspring are damaged, not enhanced. And the offspring cannot produce additional young. This four-winged fruit fly does not constitute scientific evidence in support of the theory of evolution.

But the four-winged fruit fly is still presented in most college and high-school textbooks of today as an example of how evolution does proceed, and it is taught to the students as a scientific evidence in support of the theory of evolution.

CHAPTER 23. STRUCTURAL SIMILARITY OF ANIMAL BONES

Similar Designs Suggest but do not Prove Common Ancestry

Evolutionists believe that if particular parts of several different animals are similar in geometric structure, this is scientific evidence that these animals evolved from a common ancestor. The particular example that is often cited by evolutionists is the bones of the limbs of several animals, and it is evident that they are similar in their design characteristics. The bones of the limbs of several animals are shown in Figure 23.1. A study of these bones will reveal that they are of a similar structural design. This similarity is called "homology", which word may be generally defined as "similarity of structure." But in recent years biologists have insisted that the word, "homology", in the context of biology, should be defined as "similar in structure due to common ancestry", and the dictionaries actually now so define it. But such a definition is not scientific because it *assumes* that animals have evolved from common ancestors. But, as we have proved in Chapters 12 and 13, there is no scientific evidence that evolution ever took place, or ever could have taken place, and the only historic facts available, namely the fossils, do not support the theory of evolution and its assumption of common ancestry.

Figure 23.1 Bones of Different Animals have Similar Designs

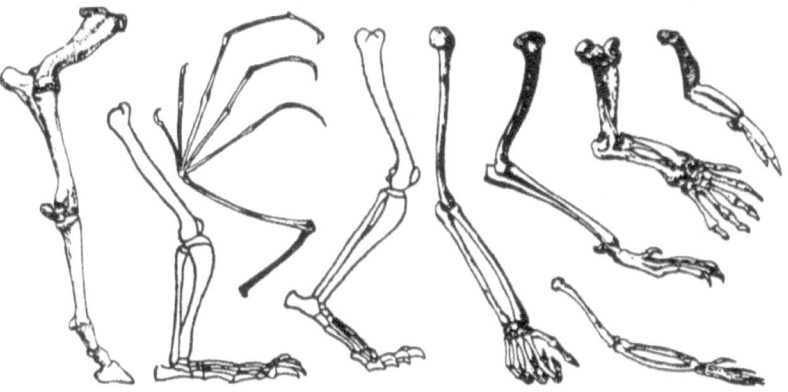

Horse Bear Bat Tiger Man Monkey Dog Manatee Penguin

It is legally and morally permissible for an evolutionist to believe that similarity of structures in animals suggests that they evolved from a common ancestor, even though they are wrong in this assumption. But for a dictionary to assume that evolution took place, and that animals evolved from common ancestors, and then to define a word, such as "homology" as "similarity of structure due to a common ancestor" is improper, unscientific and lacking in scholarliness.

The Two Explanations for Similar Designs

As we study the similar bone designs in a wide variety of animals we should attempt to explain how it happened that this similarity came about. There are two explanations which have been proposed to answer this question.

1. The animals which share similar designs might have evolved from a common ancestor.

2. An intelligent designer of animals might have recognized an excellent design, and then elected to apply this design in many animals.

Neither of the above two explanations can be proved by pointing to observations or by performing experiments. No one alive has ever observed evolution in action, or the creation of an animal by an intelligent designer, but objective analyses might reveal which explanation is more likely to be the correct one. Consider the following.

An Excellent Design Should be Used in Many Applications

In 1912 Charles Kettering invented an electric starter for an automobile engine. Cadillac adopted the starter in that year. Prior to that date automobiles were started by manually rotating a crank mounted in the front of the vehicle. Now all Ford, Chrysler, and General Motors cars have homologous starters. Did they evolve from a common ancestor or did intelligent designers recognize an excellent design and decide to use that design in a wide variety of automobiles. This example of intelligent design in action should help us understand homology in animals. The widespread usage of an excellent bone design in animal limbs could have been the brilliant application of an excellent design by an intelligent designer.

Without Evolution Biological Homology Could not Occur

If it could be proved that evolution never took place, then, no animal could have evolved from any common ancestor. In Chapter 13 we did, in fact, prove that evolution never did take place. Let's revue that proof again, so that we can fully appreciate its profound and far-reaching applicability to a wide variety of biological applications. For evolution to take place, or for one or more species to evolve from a common ancestor, it would be necessary for millions of atoms to be moved to new locations in the DNA of the common ancestor. But we showed that there is only one chance in 1.3 trillion that just 15 atoms could be successfully moved by chance. Hence it would be statistically impossible for millions of atoms to be moved by chance. In the light of this proof, the word "homology" should be removed from the biological lexicon, and we should accept the fact that similar designs did not come about by evolution from a common ancestor.

But if an evolutionist were convinced that no intelligent designer of animals ever existed, or if he believed that no intelligent designer of automobiles ever existed, then that evolutionist would be forced to assume that animals evolved from a common ancestor and that the starting motors of our cars today evolved from the one that first came into use in 1912.

Intelligent Designer Employed Excellent Design in many Applications

We must conclude then, based on the principle defined in Chapter 16, that the first explanation given above, which includes evolving from a common ancestor, cannot be the correct one. The similarity of design in various animals is due to the intelligent designer's decision to employ an excellent design in many different animals as each species was independently created from time to time during the past 543 million years.

But in spite of the obvious superiority of the intelligent-design explanation, it is a fact that the similarities of design structures in many animals is still presented in most college and high-school textbooks of today as evidence of a common ancestor, and it is taught to the students as a scientific evidence in support of the theory of evolution.

CHAPTER 24. THE MILLER-UREY EXPERIMENT

The Miller-Urey Experiment Produced Compounds Found in Living Cells

In 1953, University of Chicago professor of chemistry, Harold Urey, and his graduate student, Stanley Miller, performed an experiment which was designed to prove that a living creature could be produced by the chance actions of natural forces. They constructed an experimental apparatus which applied an electrical spark to a mixture of gases which they thought simulated the atmosphere of the early earth, specifically, methane, ammonia, hydrogen and water. This apparatus is shown in Figure 24.1 Urey and Miller assumed that the earth's early atmosphere was composed of the same chemicals as were found in interstellar gas clouds. The experiment did produce several amino acids and other organic compounds found in living cells. From this, Miller and Urey concluded that natural forces, such as lightning, acting on the early atmosphere very well could have produced the first single-cell animal.

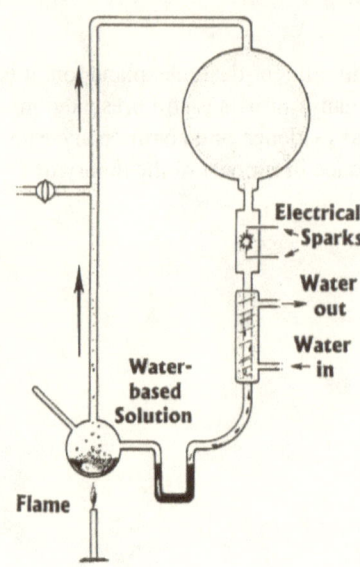

Figure 24.1 Miller-Urey Apparatus

Copyright Jody F. Sjogren 2000.
Used with permission.

Since 1953, other scientists have concluded that the atmosphere of the early earth consisted of gases produced from the earth's volcanoes, which consisted of water-vapor, carbon-dioxide, nitrogen, and some hydrogen. Therefore, the science community today has concluded that the Miller-Urey experiment proved nothing about the origin of life.

A Few Compounds are Incomparably More Simple than a Living Cell

Furthermore, and of much greater significance, are the following facts. First, it is a huge stretch of the imagination to conclude that the production of one of the organic compounds found in a living cell might suggest that the same phenomenon could, therefore, produce something as complicated as a single-cell animal. Recent research has shown that single-cell animals, are extremely complicated, comparable in complexity to a city full of machinery and chemical plants.

A Single-cell Animal has 68,000,000 atoms in its DNA

In order to produce a single-cell animal it is necessary to move about 68,000,000 atoms of O, H, N, C, and P from the land, sea and air into a very specific array of atoms to form the DNA molecule for a single-cell animal. Does the Miller-Urey experiment explain how this movement of 68,000,000 atoms could be achieved? No one has ever explained how, by chance, 68,000,000 atoms could be organized into the very specific array needed to produce the DNA molecule for a single-cell animal.

Even though the Miller-Urey experiment has for many years been rejected by knowledgeable scientists, and even though it can be shown by statistical analyses that it has no validity, it is still presented in modern biology textbooks as evidence for the concept that a first single-cell animal came into existence by the chance actions of natural forces, and it is taught to the students as scientific evidence in support of the theory of evolution.

CHAPTER 25. THE PEPPERED MOTHS

The Peppered Moths are an Example of Natural Selection

Charles Darwin believed that natural selection was the dominant process which energized evolution. And, from Darwin's day to the present, evolutionists have echoed this same refrain. Even today our teachers and textbook writers have not graduated from this simplistic dogma. It is suggested here that you read again the quotations in Reason 3 of Chapter 9. The scientists quoted there assert that natural selection is the essence of evolution. But we showed in Chapter 9 that natural selection is not a new-feature-producing agent and it has no capability to produce evolution. And, the peppered moths are a good illustration of the existence of natural selection, but the peppered moths are not an example of evolution in progress.

The Two Elements that Must be Effective for Evolution to Take Place

But, as we have explained in detail before, there are two processes which must be active and effective if evolution is to take place. They are:

1. There must be some new-feature-producing agent which is effective.

2. Natural selection must take place.

Absolutely there must be some new-feature-producing agent that can produce some new feature, and then natural selection will determine whether that new feature

Figure 25.1 A Dark-colored Moth on a Light-colored Tree. Poor Camouflage.

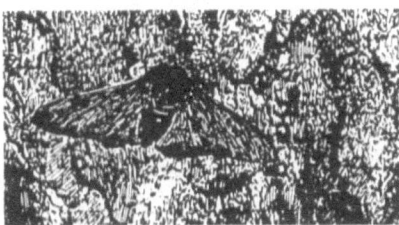

Figure 25.2 A Dark-colored Moth on a Dark-colored Tree. Good Camouflage.

Figure 25.3 A Light-colored Moth on a Dark-colored Tree. Poor Camouflage.

Copyright Jody F. Sjogren 2000. Used with permission.

is an advantage or a disadvantage. Natural selection alone does nothing other than cause an unfit animal to become extinct. Natural selection is not evolution.

The Search for Natural Selection

Nevertheless, apparently Darwin and his followers did not understand the true role of natural selection, and since they thought that natural selection produced evolution, they thought that if they could see natural selection in action, they would be observing evolution. Therefore the early evolutionists diligently searched for some visible process in nature where natural selection might be taking place.

The Peppered Moths

Accordingly, in the early 1950's British biologist Bernard Kettlewell, and others, noticed that the moth, Biston betularia, existed in two colors. Some were light-colored and others were dark-colored. Both had dots on their wings and hence they were called "peppered moths." Kettlewell observed that predatory birds ate these moths and he reasoned that in an area where the tree trunks were dark-colored, the birds could better see the light-colored moths, and the resulting predation would favor the survival of the better-camouflaged dark moths. But if the tree trunks were light-colored, the light-colored moths would be the survivors. Figure 25.1, 25.2 and 25.3 show light and dark colored peppered moths on light and dark colored tree trunks. The good and poor camouflages that result can be seen.

Research, which involved introducing thousands of moths into the woods and counting those that survived, generally supported the concept that camouflage enhances survival. And Kettlewell and his followers excitedly proclaimed that they had actually witnessed evolution in action, possibly for the first time.

Some Peppered-moth Research Problems

In 1959 Kettlewell stated that his experiments with peppered moths produced "the most striking evolutionary change ever actually witnessed in any organism" and that the results of his research provided "Darwin's missing evidence." However, since Kettlewell's first experiments, many other evolutionists have criticized his work by pointing out such facts as the following. Peppered moths in the wild don't rest on tree trunks. Moths placed on tree trunks in the daytime by researchers stay there until dark, but at night, and in the wild, the moths rest under horizontal branches. Parasites may have had an effect on the research. In England it was thought that industrial pollution darkened tree trunks, thus having an effect on the moths. But there might have been different tolerances to industrial pollution. Also, other evolutionists, eager to acquire photographs of this important discovery of evolution in action, actually pinned or glued dead moths onto tree trunks and photographed them. These faked pictures were then distributed to the unsuspecting public, and some of these fraudulent photographs have even found their way into school textbooks of today.

Natural Selection Alone does not Produce Evolution

However, the informed readers of this book will recall that natural selection is not evolution. All that natural selection does is cause the unfit animals to become extinct. None of the peppered-moth researchers has ever indicated that he is familiar with the fact that for evolution to occur there must be some effective new-feature-producing agent. Also it is significant that no researcher has claimed that any species of peppered moth has evolved into another species. The only process that these researchers seemed to be looking for was to observe natural selection in action. Also it must be remembered that no step in the process of evolution can be accomplished by any means other than by adding to or rearranging the atoms in the DNA molecule of the predecessor animal. But the peppered-moth researchers have never mentioned anything about DNA.

The Peppered Moths are not Unique Examples of Natural Selection

Furthermore, there is nothing special about observing peppered moths. Natural selection in action is occurring every day all over the world, and it is not difficult to observe. Today there are hundreds of species that are on the verge of becoming extinct. It is natural selection that is causing this process, and we can observe these in-process extinctions simply by studying our wild animals. The panthers of Florida, the condors of California and the pandas of China are near to extinction. Several hundred thousand animals have already become extinct since the Cambrian explosion. All biologists, including creationists, believe that natural selection does take place. But its role is not that of producing evolution. It is not a new-feature-producing agent.

Thus we can conclude that the peppered moths, and many other animals, do show us natural selection in action. But natural selection is not a new-feature-producing agent, and natural selection alone is not evolution. Hence the peppered moths do not in any way support the theory of evolution.

But, regardless of these facts, the peppered moths are still presented in most college and high-school textbooks of today as unique evidence that natural selection does take place in nature, and it is suggested that natural selection is synonymous with evolution. The story of the peppered moths is taught to the students as scientific evidence in support of the theory of evolution.

CHAPTER 26. THE FINCHES OF THE GALAPAGOS

Darwin did not Research the Finches

Although the finches of the Galapagos islands are sometimes called the Darwin finches, this is actually a misnomer. Darwin was aboard the British ship, the HMS Beagle, in 1835 when it visited the Galapagos, but he made no research studies of the finches of the Islands. He was not an evolutionist in 1835, and he made no mention of these finches in 1859 when he published *The Origin of Species*. But 112 years later, in 1947, David Lack published a book which he entitled *Darwin's Finches* (Lack 1). Lack believed that these finches exhibited traits of anatomy and inheritance which he thought demonstrated Darwin's theory of evolution by natural selection.

The Beak Size Correlated with Rainfall

Starting in 1973 and extending for several decades, Peter and Rosemary Grant spent much of their time studying the finches of the Galapagos. They determined that during periods of drought, the birds with large beaks flourished and those having smaller beaks diminished in numbers. Then, when the rainfall was generous, the finches with small beaks outnumbered those with large beaks. Figure 26.1 shows a finch with a large beak, and Figure 26.2 shows a finch having a small beak.

Figure 26.1 A Finch of the Galapagos Islands having a Large Beak.

Figure 26.2 A Finch of the Galapagos islands having a Small Beak

Copyright Jody F. Sjogren 2000. Used with permission.

The Grants' explanation for this was that during periods of drought, there were fewer seeds available, but the birds with large beaks were capable of cracking and eating large seeds some of which were available. But the birds with smaller and weaker beaks were not capable of cracking the large seeds. Also during droughts there were very few smaller seeds. But during years when the rainfall was abundant the numbers of small and soft seeds were also abundant. The results of the Grants' observations were greeted with great enthusiasm by evolutionists. They interpreted the correlation of beak size with rainfall as a "particularly compelling example"

of natural selection in action, and a scientific evidence in support of the theory of evolution. And almost every biological textbook of today includes the beak sizes of the finches of the Galapagos Islands as one of the most cogent proofs of evolution.

Natural Selection Alone is not Evolution; an NFPA is Needed

But there are many facts which refute the right of the Galapagos finches to be given the exalted status of a compelling example of evolution. First, it should be appreciated that the responses of the finches' beaks to weather conditions is nothing other than one of hundreds of examples of natural selection which could be cited. And, as we have stated many times before, natural selection is not evolution. We must repeat here that for evolution to take place two elements must be active and effective:

1. A new-feature-producing agent (NFPA) must be present.

2. Natural selection must take place.

Evolutionists who tout the finches never identify any new-feature-producing agent.

No New Species are Identified

Secondly, evolutionists do not clearly identify specific different species of finches, one of which has evolved from another. They merely infer that minor modifications due to natural selection must be related to the evolution of new species. This is an unwarranted extrapolation.

No Mention is Made of DNA Atoms

Thirdly, those who study the finches do not seem to be aware of the fact that for one species to evolve into another would require that thousands of atoms in the DNA molecule of a gamete in the predecessor animal must be correctly moved to different locations in the DNA of the new species, and there is no statistical probability that this could be accomplished by chance.

Maybe there was Only One Species

Furthermore, recent studies have suggested that, of the alleged 13 finch species that live in the Galapagos, many have interbred with each other and have produced fertile offspring. It has been suggested that maybe there is only one species of finch in the Galapagos. Based on the above facts, we can conclude that the Galapagos finches do not support the theory of evolution.

But, regardless of the facts, like the peppered moths, the Galapagos finches are still presented in most college and high-school textbooks of today as unique evidence that natural selection does take place in nature, and it is suggested that natural selection is synonymous with evolution. And the story of the Galapagos finches is taught to the students as scientific evidence in support of the theory of evolution.

CHAPTER 27. FACTS AND ASSUMPTIONS

Which are Facts and Which are Assumptions
The theory of evolution consists of a mixture of facts and assumptions, and in order to evaluate the validity of this theory it is necessary to distinguish between what are facts and what are assumptions. A fact is something that is known with certainty because it has been objectively verified by observation, by experimentation or by mathematical proof. An assumption is something that is assumed to be true without proof or demonstration. An assumption may be true or false.

The fact-vs-assumption distinguishings presented below will cover the topics of the previous nine Chapters. In each of the paragraphs below we will identify which statements are facts and which are nothing but assumptions, and, where applicable, we will suggest which assumptions are false, and we will explain why they are false.

The Family Tree of Animals.
FACTS: it is a *fact* proved by the fossils that different species of animals have appeared on the earth, from time to time, over the past 540 million years. Most of them lived a geologically short period of time and then became extinct. As the time has passed during this 540 million years, these animals have not become steadily larger nor more complex. The largest animals were the dinosaurs and they became extinct about 65 million years ago. The animals of the Cambrian explosion were very complex, and we are learning today that our single-cell animals also are extremely complex.

ASSUMPTIONS: The depiction of the animal kingdom as a tree on which the fruit are the animals, and the branches represent lines of evolutionary descent from early animals to later ones, is nothing but an *assumption*. There are no series of fossils which clearly show a gradual multi-step evolution from one animal to a later one. If it were true that each animal evolved from some predecessor animal there would be millions of fossils which would show half-finished new parts, and intermediate forms, but no such fossils have been found. Also, the concept that life began with the chance creation of a first single-cell animal, which might represent the root of a tree, is totally contrary to what the fossils teach us. The fossils reveal that complex life began with the Cambrian Explosion, in which dozens of very complete, involved and intricate animals suddenly appeared, with no evidence of any evolutionary past. So, the tree of animal life is nothing but an *assumption*.

Identical Early Embryos
FACTS: It is a *fact* that every animal starts out as a single cell, and then becomes two cells, three cells, etc. German biologist, Ernst Haeckel, in the mid 1800's conceived the idea that if every animal evolved from some predecessor animal, there should be some groups of animals each one of which evolved from some common ancestor. He then reasoned that such common ancestors might be

confirmed if the embryos of a particular group of animals all looked alike at some early stage of development. Accordingly, Haeckel made some drawings of the early embryos of such a group, which included fish, salamander, tortoise, chick, hog, calf, rabbit and human. But Haeckel apparently was so enthused with his idea that he exaggerated his drawing and, without really researching the matter, he drew a group of embryos for these animals with all of the embryos being *identical*. But the *fact* of the matter is that early embryos are not identical, and, in 1997, British embryologist, Michael Richardson, published actual photographs of Haeckel's animals, and it became apparent that they are not at all alike. It was widely known among biologists in Haeckel's day, and throughout the past 150 years, that Haeckel's drawings were fakes, and many scientists so stated in publications.

ASSUMPTIONS: Haeckel's drawings, and similar but more accurate depictions, are based on the *assumption* that animals evolved one from another, and that there are many groups of animals in which each one is thought to be descended by evolution from a common ancestor. It is then *assumed* that if the embryos of a certain group of animals look alike, this would prove that there was descent from a common ancestor. But there are no facts or proofs to support this *assumption*. The fossils, which are the facts, do not support the concept of a common ancestor. Also, in Chapter 13 we proved that evolutionary lineages do not exist. The concept involving early embryos is nothing but an *assumption*.

Archaeopteryx, the Missing Link

FACTS: It is a *fact* that this creature did exist. Excellent fossil remains of eight of these animals have been found. It is also a *fact* that Archaeopteryx has some features that resemble a bird, and some that resemble a reptile. Archaeopteryx suddenly appeared on earth, lived a short while and then became extinct.

ASSUMPTIONS: It is a far-fetched and reckless *assumption* to assert that birds evolved from reptiles, or that Archaeopteryx is a missing link or an intermediate creature in the animal lineage between a reptile and a bird. There are no fossils which show a gradual change from a reptile to Archaeopteryx, or from Archaeopteryx to a bird. Birds have dozens of special features that make them light enough to fly, and give them the aerodynamic mechanism which enables them to fly. No other animal has these unique features, certainly not a reptile. The assertion that Archaeopteryx is a missing link between a reptile and a bird is nothing but a farfetched *assumption*. And we should not forget that the proofs of this book show that no animal ever evolved from another.

The Evolution of the Horse

FACTS: It is a *fact* that the fossils of many animals that look something like our horse of today have been found. These animals have existed during the past 50 million years. These animals vary in size and in numbers of toes, but they are all generally similar in anatomical structure. Based on the fossils, these animals can be classified into different species and a name has been given to each species. The suggestion that the evolution of the horse could be documented with fossils from the western United States became popular in the early 1900's, and it has remained in favor even until today, but this is nothing other than a suggestion.

ASSUMPTIONS: Evolutionists have organized these horse-like animals into an array which is based on the time of existence and on anatomical characteristics. But they then *assume* that each horse evolved from some predecessor horse-like animal, and that there is an evolutionary line of descent from one of the early species, Hyracotherium, to the horse of today. Hyracotherium was a smaller animal and it had four toes. Again, there is no proof that any horse evolved from any other such animal. The suggestions of evolutionary linkage are nothing but hypothetical unproved *assumptions*. There are no series of fossils which show a gradual evolution from one horse to another. Each animal appeared suddenly, without any evolutionary background. The in-between fossils are missing. Furthermore, there is nothing special about horse-like animals. Other animals, such as deer, rabbits, cats, bears, etc. could be lined up and treated as examples of evolutionary lineages. The fossils are facts, but the evolutionary lineages, and the assertions of descent with modification are nothing but unfounded *assumptions* and *conjecture*.

The Four-winged Fruit Flies

FACTS: It is a *fact* that there are mutation-causing agents in the natural world, such as sun radiation, cosmic rays, viruses, chemicals, and errors in DNA replication. Man-made X-rays can also cause mutations. It is also a *fact* that such mutation-causing agents can produce changes in an animal. It is also a *fact* that controlled breeding of mutant fruit flies, such as, Drosophila melanogaster, did produce a fruit fly with four wings, rather than two. The extra pair of wings on the four-winged fly consisted of modifications of the balancer arms which are normally attached to the fly's thorax behind the normal set of wings.

ASSUMPTIONS: Based on a superficial acquaintance with the four-winged fruit fly, the uninformed might *assume* that the artificial production of a four-winged fruit fly from a two-winged fly would be an accomplishment that would demonstrate evolution in action. At first glance it might appear that a four-winged fly would be superior to a two-winged fly. But these *assumptions* are invalid for several reasons. First, the controlled experiments which produced the new fly would never be duplicated in the wild. Secondly, in the wild, mutations are extremely rare, and they are usually detrimental rather than advantageous. Thirdly, the four-winged fly is not better than the natural two-winged fly because the new wings had no muscles and they hindered, rather than helped, the insect to fly. Also, the new wings were modifications of, and they eliminated, the very useful vibratory balancing levers that were mounted on the thorax behind the normal wings. Mating also was hindered. The four-winged fly, if turned loose in the wild, would be eliminated by natural selection. Therefore, to assert that the production of this four-winged fly was an example of evolutionary advancement, and that it constituted scientific evidence in support of evolution, is a totally unwarranted *assumption*.

The Structural Similarity of Animal Bones

FACTS: It is a *fact* that comparable bones of different animals do have similarity. It is a *fact* that the limb bones of a horse, bear, bat, tiger, man, monkey, dog, manatee and penguin all have similar designs.

ASSUMPTIONS: To suggest that similarities in the structural designs of widely different animals is proof that each of those animals descended by evolution from a common ancestor is an irresponsible and naive *assumption*. In the first place we proved in Chapter 13 that animals do not descend one from the other by evolution. Secondly, it is logical, and to be expected, that if a designer of animals conceived a particularly good design, he would apply it in the creation of many different animals. Most automobiles have four wheels, a fossil-fueled engine, gears to transmit power, springs, and a steering wheel. Why? Because all of these are good designs. But evolutionists never mention the above obviously-credible explanation. A good design should be used repeatedly in different applications. To suggest otherwise is nothing but an unwarranted *assumption*

The Miller-Urey Experiment

FACTS: It is a *fact* that in the Miller-Urey experiment, electric sparks were applied to a mixture of water and the gases which they thought were present on the early earth. It is a *fact* that these experiments were performed and that they produced amino acids and other organic compounds that are found in living cells.

ASSUMPTIONS: To assert that the production of a few chemicals that are found in living cells should suggest that the natural forces of the universe, by chance, could have produced the first single-cell animal is a preposterous *assumption*. Each single-cell animal has in it a DNA molecule which is composed of a specific array of about 68,000,000 atoms of H, O, N, C and P. Recent research has shown that every single-cell animal is extremely complicated, containing the analogous equivalent of an entire city filled with machinery and chemical plants. It is a ridiculous and naive *assumption* to proclaim that the natural forces of the earth could, by chance, assemble 68,000,000 atoms, form a DNA molecule, and produce a single-cell animal. And the organic compounds that Miller and Urey produced were almost infinitely less complex than a single-cell animal. The suggestion that the Miller-Urey experiment illustrates the origin of any animal is certainly a farfetched *assumption*.

The Peppered Moths

FACTS: With respect to the peppered moths, it is a *fact* that the pepper-marked moth, Biston betularia, existed in two colors, dark and light. It is a *fact* that birds ate these moths as a part of their diet. It is a *fact* that any animal that is well camouflaged in its environment is less likely to be observed and eaten by a predator than one which is not well camouflaged. No mention is made by evolutionists that any species of peppered moth ever evolved into a new species.

ASSUMPTIONS: Based on the research of biologist Bernard Kettlewell, who placed peppered moths on tree trunks of different colors, and then counted those that survived predation, it has been *assumed* that, in the wild, dark moths on dark trees, and light moths on light trees, will be the survivors. And it is affirmed by most evolutionists that this observation, if true, would be an excellent example of natural selection in action. But they also *assume* that this would be evidence in support of the theory of evolution. If the better camouflaged moths did survive in greater numbers, this would be nothing other than one of many examples of natural selection in action. Natural selection does take place, and what it does is to cause unfit species to become extinct. For evolution to take place there must be some new-feature-producing agent that is effective, but natural selection is not a new-feature-producing agent. Hence to assert that any example of natural selection in action is an example of evolution taking place, is an invalid and uninformed *assumption*.

The Finches of the Galapagos

FACTS: It is probably a *fact* that birds with large beaks could better survive drought periods than birds with small beaks, and that during periods of much rainfall, the numbers of finches with small beaks might increase. No mention is made by evolutionists that any species of finch ever evolved into a different species.

ASSUMPTIONS: The finches of the Galapagos Islands is another example in which natural selection is observed, and this is *assumed* to be an illustration of evolution in action. Natural selection can be observed throughout the animal kingdom, but natural selection is not a new-feature-producing agent and it is not synonymous with evolution. Evolution cannot take place without a specific rearrangement of the atoms of the DNA molecules, and no mention is made of this taking place on the Galapagos Islands. To suggest that the Galapagos finches are an example of evolution in action is certainly an invalid and uninformed *assumption*.

Conclusions with respect to "Numerous Scientific Evidences."

When asked to provide evidence that the theory of evolution is a fact, evolutionists usually cite the "numerous scientific evidences which support the theory of evolution." And what are these evidences? They are the items covered in this Chapter, and in the previous nine Chapters of this book.

But our careful analyses of the contents of these ten chapters indicate that each of these evidences consists of a few facts and many assumptions. An assumption is not a scientific evidence. It is merely a guess. And I think we must honestly conclude that in these "evidences" the assumptions overwhelm the facts, and we can safely say that evolution really is based on hypotheses, speculations, and partisan-motivated conjecture. But in this book, especially in Chapters 12, 13, 14, and 15, we have proved with mathematical certainty that evolution could not, and never did occur, and the theory of evolution is really nothing other than one huge invalid *assumption*.

CHAPTER 28. INTRODUCTION TO INTELLIGENT DESIGN

Evolution has been Exposed

In the first 27 Chapters of this book we have exposed the theory of evolution to be what it really is, namely preposterous misinformation with which zealots have propagandized the world, and so we are now in a position to explain what is the theory of intelligent design. This theory is claimed by many scholars to be the only really truthful explanation which accurately reveals what was the origin of our animals on this earth.

Dover Pennsylvania Lawsuit

On December 20, 2005 U. S. District Judge John E. Jones released his ruling on a law case that had been in progress near Dover, Pennsylvania, throughout the autumn of 2005. In this case the Dover Area School Board had decided that the concept of intelligent design should be introduced into the science curriculum of the school, together with the usual presentation of the theory of evolution. The Judge's ruling included the following statements, which obviously must have expressed his beliefs: the theory of evolution is a science, intelligent design is not a science, intelligent design is nothing but a religious view of animal origins, the theory of evolution is not a religious view of animal origins, the introduction of the concept of intelligent design into the school curriculum would violate the constitutional separation of church and state.

These statements which were included in Judge Jone's rulings form a good group of topics that should be discussed as part of our introduction into the subject of intelligent design. To discuss these concepts it will first be necessary to define some terms.

Definition of Science?

The American Heritage Dictionary states that science is, "a domain of systematically organized and classified knowledge." The unabridged Webster dictionary states that science is, "knowledge as distinguished from ignorance or misunderstanding . . . a branch of systematized knowledge that can be made a specific object of study . . . accumulated and accepted knowledge that has been systematized and formulated with reference to the discovery of general truths." Webster then states that a synonym of science is the word, "knowledge."

From these definitions we can conclude that one of the primary characteristics of that which is science must be truth, veracity, integrity, that which is factual, and that which is known. Certainly that which is untruthful, deceptive or fraudulent cannot be defined as science. Science is basically knowledge that is truthful.

Definition of Engineering?

Many otherwise well educated professionals do not really understand the difference between a scientist and an engineer. The objective of a scientist is to discover new knowledge. The objective of an engineer is to design new products. The dictionary defines engineering as "the application of scientific principles to practical ends such as the design, construction, and operation of efficient and economical structures, equipment, and systems." Mechanical engineers design machines. Electrical engineers design electronic or electrical products or systems. Chemical engineers design chemical reactions and the equipment required to bring about these chemical reactions. Civil engineers design the supports and housings for the products designed by other engineers.

Although engineering is a design process and it is not a science, engineering is based on scientific knowledge. Engineering is applied science. The sciences of physics, chemistry, and mathematics are the underlying knowledge which engineers use to develop the general principles and formulas which are then applied to design new products. But the basic function of an engineer is to design new products.

Definition of Design?

The dictionary defines design as, "To conceive in the mind, invent, contrive or draw up plans for the details of how something is to be made, and it is the arrangement of the forms, parts, or details of something according to a plan."

Definition of Intelligence?

To arrive at the topic which is the subject of this book, we must add the word, "intelligent" to the word. "design." The word, "intelligent" means, "having intelligence", and the word, "intelligence" is defined as, "The capacity to acquire and apply knowledge and to solve problems."

Definition of Religion

The dictionary defines religion as, "The expression of, belief in, and reverence for a superhuman power, or any particular integrated system of this expression." Or "any objective pursued with zeal or conscientious devotion."

The Theory of Evolution

The theory of evolution asserts that some random, chance-based, mutation-causing agent might act upon and modify one individual animal and cause it to acquire some new feature that would enhance its ability to survive, and that this beneficial feature might then be passed on by inheritance to this animal's offspring. Several such occurrences might ultimately produce a new species. Evolutionists assume that all animals came into existence by this random mutation process. Evolutionists study existing animals, and they also study the fossils of earlier animals. They then attempt, by comparing anatomies, to determine that one particular animal must have evolved from another, or that two animals had a common ancestor.

The Theory of Intelligent Design

By combining the above definitions we can arrive at the conclusion that intelligent design is "the capacity to acquire and apply knowledge which would enable a designer to conceive in his mind, invent, or contrive plans for the details of how something is to be made or how the forms, parts, and details of something could be arranged according to a plan." Intelligent design is design that is being performed by some being that has intelligence.

Is the Theory of Evolution a Science?

The Judge in the Dover lawsuit stated that intelligent design is not a science. By so stating, and based on his ruling, it can be concluded that he thinks that the theory of evolution is a science. Therefore it might be appropriate here to study what is a science, and see if the theory of evolution qualifies as a science.

A summary of the definitions given above would assert that science is knowledge that is truthful. But what is evolution? Evolution is not knowledge, it is an engineering design process. We should recall the difference between science and engineering. The role of a scientist is to discover new knowledge. The role of an engineer is to design new products. If evolution were to take place it would be an engineering-design process the objective of which would be to design new animals. Therefore evolution is not a science, it is a design process.

The body of knowledge that describes animals and provides information concerning their way of life, such as would be presented in a textbook on zoology, would be science. It would be knowledge that is truthful. But evolution, itself, is a design process and it is not a science. Therefore to refer to evolution as a science is to reveal a lack of understanding concerning what these words mean.

Secondly, the theory of evolution is not truthful, and for this reason, also, evolution is not a science. A science must be truthful knowledge that consists of facts, not just assumptions. We proved irrefutably in Chapters 12, 13, 14, and 15 of this book that evolution never could have occurred. Therefore the theory of evolution is not truthful. We also showed in Chapters 17 through 26, and also in Chapter 27, that efforts on the part of evolutionists to support the theory of evolution are based on nothing but assumptions.
Therefore, since the theory of evolution is untruthful, and it is based on nothing other than assumptions, it is not a science, and it does not explain the origins of our animals.

Is the Theory of Intelligent Design a Science?

As explained above, intelligent design is an engineering-design process which can be applied to both machines and to animals. And for the same reasons as are given above, intelligent design is not a science. It is a design process. However, intelligent design, like all design processes, is based on scientific knowledge.

Engineering is essentially applied science. And intelligent design is unassailably truthful.

Is the Theory of Evolution a Religious View of Animal Origins?

In the Dover lawsuit the Judge ruled that intelligent design was a religious view of animal origins, and as such, its introduction into the curriculum in schools violated the constitutional separation of church and state. By inference we can conclude that the Judge believed that the theory of evolution is not a religious view of the origins of animals. Let's consider these beliefs in some detail.

The definition quoted above states that religion is, "The expression of, belief in, and reverence for a superhuman power, or any particular integrated system of this expression." Or "any objective pursued with zeal or conscientious devotion."

Certainly atheism and the theory of evolution are pursued by evolutionists with great zeal and conscientious devotion. Therefore, according to that part of the definition, the theory of evolution is a religion. But we should also ask if the evolutionists have any beliefs that are related to a superhuman power.

As we mentioned in Chapter 9, a total of 93% of the biological and physical-science members of the National Academy of Sciences who responded to the survey in 1998 were either atheists or agnostics, and based on my studies, probably almost 100% of biological scientists are atheists. And, based on definitions, atheism is just as much a religion as theism is a religion. Atheism has to do with an attitude relative to a superhuman power. Furthermore, it is this religious belief in atheism that dominates the thinking of evolutionists on the subject of evolution. They cannot believe that any superhuman being ever existed. Therefore they must reject two of the three designing entities which are described in Chapter 8. They believe that human beings are not intelligent enough to design an animal, and no superhuman being ever existed. They are then left with the remaining entity, the natural forces of the earth. The important concept here is that it is the scientist's belief in the religion of atheism that dominates their system of beliefs.

Based on the above analyses we must conclude that the theory of evolution is a religious view of animal origins, and as such, the Judge should not permit the theory of evolution to be taught to students in the classroom.

Is the Theory of Intelligent Design a Religious View of Animal Origins?

To answer this question, let me mention a few facts. I will challenge anyone to find any statement in this book that could be construed as a religious statement. I have not included in this book any quotation from the Bible or any other religious book. I have not included in this book any quotation on the subject of religion from any scholar that could be construed as a religious person.

I have, in this book, asserted that there must exist some superhuman being who has more intelligence than any human being. I have called this being a superhuman being. But this is not a religious statement. I do not attempt to identify such a being, give him a name, or attribute to this superhuman being any characteristics other than having enough intelligence and knowledge to design and construct animals. And my belief in the existence of this superhuman being is nothing other than a logical conclusion which must be drawn from the following facts: (1) we see animals on this earth, (2) these animals had to have been designed, (3) no human being has enough intelligence to design these animals, (4) a designer with more intelligence must exist, and (5) I call this being that must exist a superhuman intelligent designer. My conclusion that a superhuman intelligent designer must exist is based on facts and logic, and not on any quotation from any religious book.

This book is based on facts, logic, and engineering analyses, not on any religious quotations, statements or doctrine.

The Supreme Importance of Design as it Relates to Machines and Animals
It should be obvious to any reasonable person that both man-made machines and animals are so complex that they could not have come into existence without being designed. There had to have been behind the appearance on earth of every machine or animal "a being who could conceive in his mind, invent, or contrive plans for the details concerning how the machine or animal is to be made." The process by which an animal comes into existence is engineering design.

The process of engineering design is the concept with which we must become familiar in order to understand the origins of the animals. The question that we must answer is which theory best describes the particular engineering-design process by which animals were designed and constructed on this earth.

Were the Animals Created by the Theory of Evolution?
The theory of evolution is based on several statements, beliefs, and assumptions, including those listed below.

(1) Evolutionists, due to their atheistic religion, do not believe that any superhuman being ever existed.

(2) Therefore evolutionists do not believe that the animals were created by any superhuman intelligent designer.

(3) Evolutionists agree that no human being is intelligent enough to design an animal.

(4) Having eliminated from consideration two of the three possible designing entities discussed in Chapter 8 of this book, the only remaining entity is "the natural forces of the earth."

(5) Evolutionists believe that the animals of the earth were created by the chance actions of the natural forces of the earth.

(6) Evolutionists ignore the fact that the natural forces of the earth have no intelligence.

(7) Evolutionists ignore the fact that the only possible way that one animal could evolve into another would be by some action that would rearrange the DNA atoms of the predecessor animal into the precise array of atoms required to produce the new species.

(8) Evolutionists ignore the fact that, based on statistical analyses, no mutation-causing agent could find the gamete cell.

(9) Evolutionists ignore the fact that there is only one chance in 1.3 trillion that just 15 DNA atoms could be properly rearranged by chance.

(10) Evolutionists ignore the fact that the fossils do not support the theory of evolution.

(11) Evolutionists ignore the fact that it is almost impossible to achieve an objective which is complex if the achieving effort is based largely on chance.

(12) Evolutionists ignore the fact that all of their often-cited scientific proofs which allegedly support the theory of evolution and which are discussed in Chapters 17 through 27, are based on assumptions rather than facts.

(13) Evolutionists rely on such concepts as natural-selection, adaptation to an environment, and group isolations, all of which have been shown to be ineffective as producers of evolution, because they have nothing to do with the DNA molecules.

What evolutionists do believe, and do zealously promote, is the theory that the animals of the earth all came into existence by the accumulation of many small modifications to predecessor animals all of which were caused by the chance actions of the natural forces of the earth. But, the evolutionists ignore the fact that the natural forces of the earth have no intelligence, and they also ignore the negative effects of the other statements and assumptions listed above. Taking all of these factors into consideration we have to conclude that the animals of the earth were not created by the theory of evolution.

Were the Animals Created by Intelligent Design?

The theory of intelligent design is based on several eminently defensible statements and assumptions:

(1) It is assumed that human beings are sufficiently intelligent to design machines, but they are not intelligent enough to design an animal.

(2) Animals, like all machines, must have been created by the process of engineering design.

(3) The process of engineering design requires an intelligent designer, who has both intelligence and knowledge.

(4) Since no human being is intelligent enough to design an animal, the intelligent designer who is capable of designing an animal must have had superhuman intelligence.

(5) Since there are millions of animals on earth, and we can see them, we have to conclude that they must have come into existence by the actions of the only being who is intelligent enough to design them, namely a superhuman intelligent designer.

Based on the irrefutable logic and truthfulness of the above statements and assumptions we can confidently conclude that the animals of the earth were in fact created by a superhuman being who had superhuman intelligence and knowledge.

The Rulings of Judge John E. Jones

As recorded at the beginning of this Chapter, Judge John E. Jones ruled that intelligent design was not a science; that it was actually a religious view of animal origins; and its introduction into the curriculum violated the constitutional separation of church and state. This judicial ruling reveals the dominant extent to which the atheistic biological scientists of the world have propagandized the other scientists, the philosophers, the judiciary, attorneys, the media and much of the general public on the subject of the origins of the animals. But in this Chapter we presented cogent arguments in opposition to the opinions of the Judge. We showed that evolution is not a science. We showed that, actually, both evolution and intelligent design were really engineering-design processes, not sciences. We showed that the theory of evolution is a religious view of animal origins. And we emphasized the fact that no mention of any religious doctrine is included in this book. And, finally, we supported with many statements the conclusion that the animals of the earth were designed by the actions of a superhuman intelligent designer, and not by the theory of evolution.

Were the Animals Created from Scratch or by Modifying Predecessors?

One interesting question has occurred to me as I have studied this subject. Did the superhuman intelligent designer create the animals from scratch, or were some of them created by modifying a predecessor animal? I think that they could have been created by either process, but I believe that there is some evidence that creation from scratch was the preferred method. The evidence in support of this opinion is contained in the records of the fossils. In the first place, all of the animals of the

Cambrian Explosion came into existence within an instant of geological time. They had to have been created from scratch. Secondly, the fossils show that each species of animal, typically, came into existence suddenly, it stayed on earth often for a long period of time, and then it suddenly disappeared. The fossils do not reveal that gradual, multiple, modifications of predecessor animals was the process by which animals came into existence on this earth. The evidence seems to indicate that, more probably, the animals were designed and constructed from scratch.

It is interesting that the evolutionists have so indoctrinated the public on the concept that animals evolved one from another that very few students have ever thought of the concept of creation from scratch.

CHAPTER 29. INTELLIGENT DESIGN IS PHILOSOPHICAL LOGIC

The Essence of Intelligent Design

The essence of the book, *Evolution Exposed and Intelligent Design Explained*, by Walter L. Starkey, can be distilled into a small group of sentences that constitute an example of philosophical logic. These sentences are recorded below.

1. It takes considerable intelligence to design any complex entity such as a machine or an animal.

2. Only living beings have intelligence.

3. Human beings are intelligent enough to design machines.

4. Animals are much more complex than machines.

5. No human being is intelligent enough to design an animal.

6. Animals exist and therefore they had to have been designed by some being.

7. The animals were designed by some being which had more intelligence than that of a human being.

8. The animals were designed by some superhuman being who had superhuman intelligence.

CHAPTER 30. SUMMARY AND CONCLUSIONS

In this book, we first identified the three entities on the earth which obviously have been designed by some design-capable agent, namely: (1) the surface of the earth, (2) machines, and (3) animals.

We then identified the three allegedly design-capable entities that are on the earth, namely: (1) the natural forces of the earth, (2) human beings, and (3) a superhuman being.

We then made three assertions: (1) the natural forces of the earth have no intelligence whatsoever, and therefore they could not have designed the animals, (2) human beings are intelligent enough to design machines but they are not intelligent enough to design the animals, and (3) the only entity that could design the animals would be the superhuman being who is much more intelligent than any human being.

We then discussed the reasons why evolutionists believe as they do, including the role of atheism in their thinking.

We then discussed the DNA molecule and mutations because an understanding of these items is crucial to any study of the origins of the animals.

We then presented four proofs each of which proved absolutely that evolution has never occurred on the earth. These included: (1) the gamete cell cannot be found, (2) there is only one chance in 1.3 trillion that just 15 DNA atoms could be properly rearranged by chance, (3) no partially-finished new features can be found on any animal, and (4) the fossils show that all of the expected intermediate steps of evolution between species are missing.

We then demonstrated the basic nature of chance, and we showed that evolution, which is based entirely on chance actions, could never have designed any of our animals.

We then studied each of the nine examples which are commonly cited as scientific proofs that support the theory of evolution, and we showed that each one of them is based on assumptions, not on facts, and if carefully analyzed, none of them supports the theory of evolution.

We devoted the first 27 Chapters of the book to the above items. These 27 Chapters exposed the theory of evolution to be a false and fraudulent explanation for the origins of the animals of the earth. The first statement of the title to this book is "Evolution Exposed." These Chapters have exposed evolution!

The final Chapter of the book fulfills the second statement of the title of this book. It consists of, "Intelligent Design Explained." It explains the theory of intelligent design.

And, since this Chapter gives our conclusions, we probably should state again here just what is the theory of intelligent design. There are millions of animals on the earth and we can see them and study them. We conclude from this study that the animals are extremely complex, much more complex than the machines that have been designed by human beings. Both machines and animals are so complex that we must conclude that they were designed by some intelligent being who is a capable designer. Animals had to have been designed by someone. Human beings are sufficiently intelligent to design machines, but they are not intelligent enough to design the animals. Therefore we must conclude that there must exist some superhuman being who is intelligent enough to design the animals. Otherwise they would not be on the earth. This superhuman being could be called a superhuman intelligent designer. The theory of intelligent design is then, obviously, a theory that asserts that this superhuman intelligent designer does exist and that he is the being that designed all of our animals. The words, "intelligent design," simply identify a designer that is intelligent. But since it is referring to animals, the intelligence of this superhuman being is much greater than the intelligence of a human being.

The alternative to the theory of intelligent design is the theory of evolution. This theory asserts that no superhuman intelligent designer exists, because such existence would be contrary to the religion of atheism. Most evolutionists adhere to the religion of atheism. Then, since evolutionists agree that human beings are not intelligent enough to design an animal, the only remaining designing entity available to the evolutionists is the natural forces of the earth. And we showed in Chapter 8 that the natural forces of the earth have no intelligence and they act strictly on the basis of chance. Therefore we have to conclude that the evolutionists base their theory on a designing entity that has no intelligence and which acts solely on the basis of chance.

Of the two theories, it seems obvious to me that the animals of the earth were designed and constructed by a superhuman intelligent designer, and the theory of evolution is the greatest scientific mistake of all time.

Logic dictates that the animals of the earth were designed by some superhuman being.

BIBLIOGRAPHY

Behe 1: *Darwin's Black Box,* Michael J. Behe
Boucot 1: *Evolution and Extinction Rate Controls*, Arthur J. Boucot, 1975, p. 196.
Coffin 1: *Liberty*, Sept. 1975, Harold G. Coffin, p. 12.
Craig 1: *Prehistoric Facts*, Annabel Craig, 1986, p.31. All Craig references are reproduced from *The Usborne Book of Prehistoric Facts* by permission of Usborne Publishing, 83-85 Saffron Hill, London, ECIN 8RT, UK. Copyright 1987 Usborne Publishing Ltd.
Craig 2: *Ibid*, p. 45.
Darwin 1: *The Origin of Species,* Charles Darwin, 1859, pp. 83, 88, 91, 92,
Darwin 2: *ibid*
Godfrey 1: *Scientists Confront Creationism,* Laurie R. Godfrey, 1983, p.158.
Godfrey 2: *Ibid*, p.158.
Gould 1: *Paleobiography,* Winter 1980, Stephen Gould.
Gould 2: *Natural History*, May 1977, Stephen Gould, p. 14.
Hartwig 1: *Moody,* Vol. 95, No. 9, 1955, Mark Hartwig, p. 16.
Hartwig 2: *Ibid*, p. 16.
Hickman 1: *Biology of Animals*, Hickman, Roberts and Hickman, pp.19, 21, Copyright 1986 by Times Mirror/Mosby College Publishing. All Hickman references are reproduced with permission of The McGraw-Hill Companies.
Hickman 2: *Ibid*, p. 463.
Hickman 3: *Ibid*, p. 257.
Hickman 4: *Ibid*, p. 24.
Hickman 5: *Ibid*, p. G-12.
Hickman 6: *Ibid*, p. 89.
Hickman 7: *Ibid*, p. 98.
Hickman 8: *Ibid*, p. 112.
Jarvik 1: See Hickman, p. 549.
Kier 1: *New Scientist*, January 15, 1981, Porter Kier, p. 129.
Kitts 1: *Evolution,* vol. 28, 1974, David B. Kitts, p. 467.
Levine 1: *U. S. News,* Vol. 123, No. 7, M. Levine via S. Brownlee, p. 76.
Luria 1: *A View of Life,* Luria, Gould and Singer, 1981, p. 638.
Luria 2: *Ibid*, p. 651.
Marshek 1: *Design of Machine and Structural Parts,* Kurt Marshek, 1987.
Morris 1: *Time*, Vol. 146, No. 23, 1995, S. Conway Morris, p.70.
Nash 1: *Time,* Vol. 146, No. 23, 1995, J. Madeleine Nash, p. 68.
The Nash references are reproduced with the permission of TIME magazine.
Nash 2; *ibid*, p. 70.
Patterson 1: *Darwin's Enigma*, Luther Sunderland, 1984, p. 89.
Raff 1: *Time,* Vol. 146, No. 23, 1995, R. Raff via Madeleine Nash, p. 74.
Raup 1: *Field Museum of Natural History Bulletin*, January 1979, David M. Raup, p. 23.
Raup 2: *Ibid*, p. 22

Ridley 1: *Evolution*, Mark Ridley, 1993, p. 5.
Ridley 2: *Ibid*, p.5.
Ridley 3: *Ibid*, p. *520*.
Romer 1: *Natural History*, Oct. 1959, Alfred Romer, p. 466.
Ross 1: *The Creator and The Cosmos,* Hugh Ross, 1993, p. 79.
Ross 2: *Ibid*, p. 118.
Ross 3: *Creation and Time,* Hugh Ross, 1994, p. 151.
Schindewolf 1: *Basic Questions in Paleontology*, Otto Schindewolf, pp. 311, 312, 343. Published by, and permissions granted by, the University of Chicago Press.
Schindewolf 2: *Ibid*, pp. 332-333.
Schindewolf 3: *Ibid*, p. 360.
Schindewolf 4: *Ibid,* pp. 168, 214.
Schindewolf 5: *Ibid*, pp. 186, 187.
Schindewolf 6: *Ibid*, p. 103.
Schindewolf 7: *Ibid, p. 16.*
Schindewolf 8: *Ibid,* p. 15.
Schindewolf 9: *Ibid,* pp. 215-216.
Schroeder 1: *Genesis and the Big Bang,* Gerald L. Schroeder, 1990.
Simpson 1: *The Major Features of Evolution*, George Gaylord Simpson, 1953, p. 360.
Sjogren 1: Illustrator for *Icons of Evolution* by Jonathan Wells
Stanley 1: *The New Evolutionary Timetable*, Steven M. Stanley, 1982, p. XV.
Starkey 1: *The Cambrain Explosion,* Walter L. Starkey
Starkey 2; *ibid*
Starr 1: *Biology Concepts and Applications*, Cecie Starr, 1991, p. 148, Copyright 1990. All Starr references are reproduced with the permission of Brooks-Cole Publishing Company, a division of International Thomson Publishing, Inc. All rights reserved.
Starr 2: *Ibid*, p. 179.
Starr 3: *Ibid*, p. 181.
Starr 4: *Ibid*, p. 180.
Starr 5: *Ibid*, p. 7.
Starr 6: *Ibid*, p. 179.
Starr 7: *Ibid*, p. 200.
Starr 8: *Ibid*, p. 202.
Wells 1: *Icons of Evolution*, Jonathan Wells

www.ingramcontent.com/pod-product-compliance
Lightning Source LLC
Chambersburg PA
CBHW022101170526
45157CB00004B/1426